Great Lakes Oil Spill Response Capabilities Evaluation

AARON C. DAVENPORT, KATHERINE ANANIA, SUSAN A. RESETAR, ALEXIS LEVEDAHL, LIAM MCLANE, JO CAULKINS, MELISSA BAUMAN

Prepared for the U.S. Department of Homeland Security, U.S. Coast Guard, Office of Marine Environmental Response Policy

HSOAC
HOMELAND SECURITY
OPERATIONAL ANALYSIS CENTER

This research was published in 2024.
Approved for public release; distribution is unlimited.

About This Report

The purpose of this report is to provide an assessment of the ability of government agencies and the private sector to collectively respond to spill scenarios relating to Area Contingency Plans, characterizing capabilities and gaps.[1] This project, U.S. Coast Guard Great Lakes Oil Spill Response Capabilities, Capacity, and Gaps Study, addresses the scope of current oil spill risks in the region; the roles, capabilities, and capacities of oil-spill response partners; and any gaps and potential mitigation strategies around oil-spill response in fresh water and ice. The findings should be of interest to oil spill responders from many different agencies, contractors, and private industry in the Great Lakes region on both sides of the U.S.-Canadian border.

This research was sponsored by the U.S. Coast Guard Office of Marine Environmental Response Policy, Great Lakes Oil Spill Center of Expertise and conducted in the Infrastructure, Immigration, & Security Operations Program of the RAND Homeland Security Research Division (HSRD), which operates the Homeland Security Operational Analysis Center (HSOAC).[2]

About the Homeland Security Operational Analysis Center

The Homeland Security Act of 2002 (Public Law 107-296, § 305, as codified at 6 U.S.C. § 185) authorizes the Secretary of Homeland Security, acting through the Under Secretary for Science and Technology, to establish one or more federally funded research and development centers (FFRDCs) to provide independent analysis of homeland security issues. The RAND Corporation operates HSOAC as an FFRDC for the U.S. Department of Homeland Security (DHS) under contract 70RSAT22D900000001.

The HSOAC FFRDC provides the government with independent and objective analyses and advice in core areas important to the department in support of policy development, decisionmaking, alternative approaches, and new ideas on issues of significance. HSOAC also works with and supports other federal, state, local, tribal, and public- and private-sector organizations that make up the homeland security enterprise. HSOAC's research is undertaken

---

[1] Project Management Plan, Great Lakes Oil-Spill Response Capabilities, Capacity, and Gaps, Homeland Security Operational Analysis Center, Infrastructure, Immigration and Security Operations Program, July 14, 2023;

Great Lakes Oil-Spill Response Capabilities, Capacity, and Gaps, Federally Funded Research and Development Center, Technical Execution Plan, Great Lakes Oil-Spill Response Capabilities, Capacity, and Gaps, U.S. Coast Guard Office of Marine Environmental Response Policy (CG-MER), Great Lakes Oil Spill Center of Expertise (GL COE), August 11, 2022

[2] National Oceanic and Atmospheric Administration (NOAA), Office of Response and Restoration, "Great Lakes Coast Guard Oil Spill Center of Expertise Grand Opening," webpage, August 29, 2022.

by mutual consent with DHS and organized as a set of discrete tasks. This report presents the results of research and analysis conducted under 70Z02322FMER00002, Great Lakes Oil Spill Response Capabilities, Capacity, and Gaps. The results presented in this report do not necessarily reflect official DHS opinion or policy. For more information on HSRD, see www.rand.org/hsrd. For more information on this publication, see www.rand.org/t/RRA2375-1.

Acknowledgments

The authors would like to thank the many subject-matter experts from Canada and the U.S., who offered valuable information regarding oil-spill response in the Great Lakes region. Special thanks to our U.S. Coast Guard sponsors: LCDR Clifton Graham, LCDR Christopher Pace, and Jerome Popiel. We greatly appreciate our quality assurance peer reviewers, Scott Savitz and VADM Robert Parker, USCG (ret), for their thoughtful review.

Summary

The Great Lakes are a source of water for 40 million Americans and Canadians, an important recreational area, and a major shipping thoroughfare for commodities and other goods. They are also potentially becoming more vulnerable to spills because of increased domestic production of oil and petroleum products in the region.

To ensure that the region's responders are well-prepared to tackle any potential spills in the Great Lakes, the U.S. Coast Guard (USCG) Office of Marine Environmental Response Policy asked RAND's Homeland Security Operational Analysis Center to study the scope of current oil spill risks in the region; the roles, capabilities, and capacities of oil-spill response partners; and any gaps and potential mitigation strategies around oil-spill response in fresh water and ice.

Issue

Approximately 1,300 spills of oil and petroleum products have occurred in the Great Lakes region between 2013 and 2019, according to data from the U.S. National Response Center, varying in volume up to 1,500 gallons. Although spills in the Great Lakes are relatively small and infrequent compared to, for example, the Gulf of Mexico and many of the other coastal USCG Districts, the risk is on the rise because of increased domestic production. The lakes' enclosed environment means that oil and petroleum products take years to dissipate and its unique conditions—ice, many types of shoreline, ecologically sensitive areas, and a fresh water environment that could require different tools from the more common salt water coastal spills—can hinder the response. Also complicating the response to oil spills is the fact that the Great Lakes encompass eight states, two Canadian provinces, and more than 100 tribal nations.

Approach

The analysis drew primarily on documentation provided by the U.S. and Canadian governments, Great Lakes response organizations, publicly available relevant research, government sponsored data, and subject-matter expert interviews. These materials characterize (1) the quantities of oil moving through the Great Lakes region and by what means; (2) any indications of potential large-scale changes in the next few decades; (3) the roles and responsibilities of the entities responsible for oil-spill response; (4) what capabilities and capacities the different entities on both sides of the border have to respond to spills; and (5) what new technologies are emerging. To assess gaps in response capabilities and capacities, we analyzed scenarios that stakeholders have used to practice and assess oil-spill response in the past. We also built a framework that responders can use to create new scenarios to practice capabilities that they have not tested before. Because of the limited amount of publicly available oil-spill response capability documentation, this research relied heavily on what was provided by the government sponsor, U.S. and Canadian oil-spill response organizations, various SMEs, and Canadian governmental organizations. Oil-spill response policy, techniques, tactics, and procedures are also present in the various area oil response plans and regulation, which we reviewed.

Key Challenges for Oil-Spill Response

Situational/Contextual Gaps

- **Cold weather conditions and fresh water ice present challenges to oil-spill detection, containment, and recovery; vessel movement; and personnel safety.** There are techniques for oil containment and recovery in Great Lakes ice conditions, but not all stakeholders have full awareness of and experience with them.
- **Modeling the Great Lakes system is especially complicated compared with more open bodies of water.** In general, "spills that affect the connecting rivers, canals, and channels are more complicated to model, and robust system-wide models have not been developed," according to the Great Lakes Commission.

Systemic Challenges

- **Operational experience with spills is limited because major spills are rare.** Personnel turnover also limits how much knowledge transfer occurs. Modeling is a systemic challenge. There are hydrological oil transport models for projecting the trajectory of an oil spill but no formal process to validate them. Additionally, salt water-based research dominates the field; how oil behaves in salt water might not be true for fresh water.
- **Coordinating response operations across agencies could be broader.** Other agencies might have needed technical expertise, but such factors as a lack resources could prevent them from quickly responding. Also, the USCG's district and sector structure is not well aligned with the other federal agencies' regional offices, which requires additional coordination.
- **Validation of Oil Spill Response Organization capabilities and capacities are limited.** U.S. and Canadian organizations have different worst-case scenario requirements for their response plans. Because capabilities for spills above 10,000 tons are not regulated, Transport Canada has no way to know if they are truly prepared for larger spills, relying heavily on collaboration between response organizations and cascading resources for such events.
- **Equipment positioning and capabilities in remote areas are challenging.** There is concern about remote areas and the ability to maintain distributed equipment caches stored in trailers throughout the area of responsibility (AOR). In remote AORs, the USCG has supported spill response and might be better equipped than its Canadian counterpart. In addition, because of the lack of boom interoperability between the U.S. and Canada, the Canadian Coast Guard's capabilities are not force multipliers for USCG capabilities (unlike the USCG National Strike Force, which has highly trained personnel and specialized equipment that is interoperable with U.S. response organizations).

In terms of the types of equipment that could be used; surface, aerial, and submersible unmanned systems could improve detection, tracking, and assessing a spill; particularly in more remote areas. There is continuing research to improve data acquisition, transfer, processing, and storage and mapping conditions in winter (including with ice) for remotely sensed information.

- **Barriers hinder access to new technologies for dealing with oil spills in fresh water.** There are policy, resourcing, stakeholder, and market disincentives to innovate. Response requirement standards are capability-driven, not performance-driven, and response is limited by law to mechanical means unless exceptional conditions apply. In general, there is some noted reluctance by response organizations to champion innovation or challenge the status quo because of stakeholder beliefs around nonmechanical means, even though under certain circumstances they can be both more efficient and relatively safe for the environment. More proactive collaboration and information-sharing across the response enterprise, specifically about advances in clean-up technology, could alleviate the trepidation with using nonmechanical techniques and the negative perceptions of dispersant use and in-situ burning.
- **Approaches to assessing risks and vulnerabilities are inconsistent.** Ways of assessing risks and vulnerabilities in the Area Contingency Plans (ACPs) were seen as inconsistent, and levels of integration between the ACPs and the Geographic Response Plans (GRPs) or Geographic Response Strategies differ. ACPs were mixed on the inclusion of water intakes, which can vary by type in terms of how the water is drawn and the distance from the shoreline. Some respondents saw the ACP as too general to be useful and relied on the GRP instead. There are also unknown risks associated with sunken vessels, shoreside underground storage tanks, and aging infrastructure, especially because the locations, type, and quantity of pollutants, condition, and other factors have not been identified.
- **Information is lacking about certain types of oil.** Although there are methods for dealing with nonfloating oils, response operations for spills of these types of oil are complex, and not all response organizations are equipped to handle them. Also, the behavior of nonfloating oils in fresh water is not well-understood because most spills occur in salt water.

Key Findings

Overall, response capabilities and capacity are robust, and risks are being managed in accordance with existing regulations and policy.

There is a high level of cooperation and willingness to collaborate, but some agencies are siloed and could improve their external communications and information-sharing, and their contributions to educating concerned citizens and community leaders about oil-spill response tactics, techniques, and procedures (TTP).

Potential cultural divergence among key stakeholders might be overcome by more collaboration and information-sharing, which is expected to improve as the Great Lakes Oil Spill Center of Expertise grows into its new role as a regional leader and builds its reputation as the go to organization for scientific SMEs, a key facilitator for translational research, and a national repository of knowledge. An example of potential cultural divergence is the perception that the

USCG naturally defaults to coastal salt water environment response TTP, which surfaces in minor but potentially significant ways: overemphasizing vessel spills in exercises, excluding drinking water contacts from emergency call lists, and focusing on response time requirements and mechanical approaches as opposed to the efficacy of the method.

Collectively, the Great Lakes response community appears to have clear roles and responsibilities. There are gaps in what their existing scenarios cover, and developing scenarios more systematically and collaboratively could uncover undiscovered gaps in both capability and capacity.

Contents

Figures and Tables

Figures

Tables

Chapter 1. Introduction

The Great Lakes are a source of water for tens of millions of Americans and Canadians, an important recreational environment, and a major shipping thoroughfare for commodities and other goods. Despite the lakes' collective size, their relatively enclosed environment makes them particularly vulnerable to oil spills. Dilution can take far longer than in the more open Gulf of Mexico, where most substantial U.S. oil spills occur. Cleanup challenges in the Great Lakes are fundamentally different from salt water spills because of the seasonal presence of ice, the lesser density of fresh water compared to salt water,[3] the presence of water intakes, numerous chokepoints, elongated and craggy coastlines, and the need for binational cooperation. The Great Lakes are also unique in many other ways that complicate response to oil spills.

To ensure that the region is well-prepared to respond to possible oil spills in the Great Lakes, the U.S. Coast Guard (USCG) Office of Marine Environmental Response Policy asked the Homeland Security Operational Analysis Center (HSOAC) to characterize the response capabilities and capacity of the agencies that are responsible for responding to oil spills and to identify any gaps in capabilities and capacity. As part of the gap analysis, we also created a framework that will help the responsible agencies (1) develop realistic and more relevant scenarios that they can use to practice their response, and (2) use the scenarios to identify current and future gaps in response preparedness on their own.

Background

Although oil spills are relatively small in number and size in the Great Lakes, the risk endures. There were 1,300 spills between 2013 and 2019, an average of approximately 180 spills a year. Figure 1.1 provides a visual of oil and petroleum product spills during that period from the U.S. National Response Center data. Spills that occurred on land but reached the Lakes were also included. The size of the spill was not consistently reported, but large spills were far less frequent than small spills. Spill volumes were reported up to 1,500 gallons. The map in Figure 1.1 documents where U.S. oil spills have occurred, but this is not predictive of the future. Additional data analysis and maps that illustrate where there are sources that have potential for future spills are presented in Chapter 2.

[3] Salt water is approximately 2.5 percent denser than fresh water.

Figure 1.1. Map of U.S. Oil and Petroleum Product Spills in the Great Lakes, 2013–2019

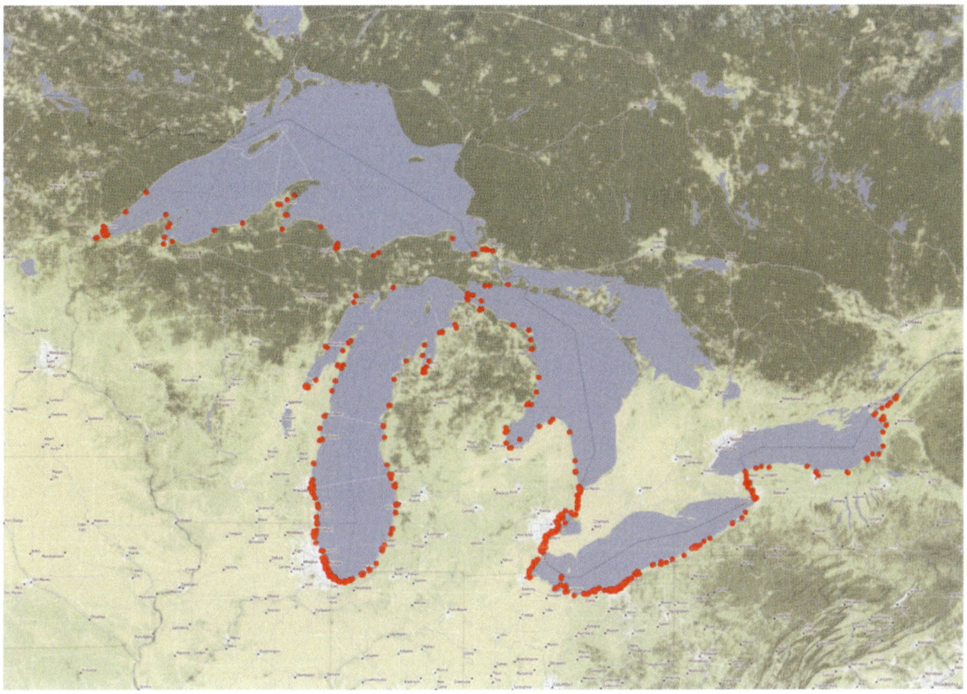

SOURCE: Features information from U.S. Environmental Protection Agency (EPA), National Response Center, database, undated.
NOTE: Figure does not include historical spills in exclusively Canadian territory.

The risk of oil spills in the Great Lakes is increasing because of increased production, as the Great Lakes Commission points out:

> Although crude oil is not routinely transported by vessel on the Great Lakes themselves, as the production of domestic crude oil continues to grow, there is an overall increase of crude oil transportation on the broader Great Lakes basin waterway system, which includes the inland waterways, rivers, and canals adjacent to the Great Lakes. The U.S. Coast Guard notes that even without factoring in cargos, the large foreign, Canadian and U.S. vessels traversing the Great Lakes and the St. Lawrence River can carry thousands of gallons of diesel or bunker C oil as fuel. The majority of the fuel oil carried on U.S. and Canadian lakes is located in internal tanks away from the skin or outer hull of the vessel, but the vessels themselves do bring a certain level of risk with them on their own.[4]

The type of response required to tackle spills depends on several factors that will be discussed in this report. One is the type of oil involved; for example, the thick bitumen extracted in Alberta, Canada, has very different properties than lighter grades of crude oil and requires different approaches to cleanup. The volume of oil to be cleaned up also clearly matters, as does

---

[4] Great Lakes Commission, *Issues and Trends Surrounding the Movement of Crude Oil in the Great Lakes–St. Lawrence River Region*, February 2015.

the spill's proximity to shore, prevailing local winds and currents, and its distance from potential response capabilities. The time of year also profoundly affects the utility of some capabilities and how others need to be employed: cold-weather conditions can take a toll on personnel and equipment, the number of hours of daylight might affect the speed of operations, and icy or heavy snow conditions can severely limit what types of response capabilities are useful and effective.

The Uniqueness of the Great Lakes

The Great Lakes region has many unique features that can complicate the effects of an oil spill and the cleanup, including environmental shifts,[5] the flow of water from one lake to another, how climate change affects the shipping industry, and sensitivities to pollutants. Potential climate change implications are presented in greater detail later in this section.

The winter 2022 edition of the USCG's *Proceedings* journal dedicated an entire edition to the Great Lakes and the region's unique operating environment.[6] The USCG Ninth District Commander offered a perspective in the forward, pointing out the "dramatic shift" in the environment because of the cold climate and ice-covered waters for about half the year, depending on location and annual variation.[7] Other highlights included the predominant maritime transportation of bulk raw materials that supply the region's manufacturing industries that are economically important to both Canada and the United States. The Great Lakes cover more than 95,000 square miles, eight states, and two Canadian provinces, and contain more than 100 tribal nations—all sharing an international land border of approximately 1,500 miles between the U.S. and Canada.[8]

Several other contributing authors to the journal representing nongovernmental organizations; private consultancy; and federal, state, or local governmental interagency provided additional detail about the unique nature of the Great Lakes. The Great Lakes contain more than 6 quadrillion gallons of fresh water—almost 20 percent of the world's surface supply of fresh water. Approximately 35 million people live in the Great Lakes watershed. The lakes provide drinking water for 40 million people, and 56 billion gallons of water are used daily in

---

[5] *Environmental shift* refers to a change or disturbance of the environment. It can be caused by human influences and natural ecological processes (Kate Wheeling, "What Causes Ecological Shifts?" Eos.org, October 3, 2017).

[6] Proceedings, *The Great Lakes, The Ninth District's Unique Operating Environment,* Vol. 79, No. 3, Winter 2022. *Proceedings* is a journal of safety and security at sea that publishes relevant articles submitted by diverse public and private interests in the maritime community three times a year. The contributing authors are routinely subject matter experts from the Coast Guard, the marine industry, interagency, or academia. The views expressed by the authors are not necessarily representative of USCG policy or service leadership.

[7] Ice is common from December to May (J. Wang, J. Kessler, F. Hang, H. Hu, A.H. Clites, and P. Chu, "Great Lakes Ice Climatology Update of Winters 2012–2017: Seasonal Cycle, Interannual Variability, Decadal Variability, and Trend for the Period 1973–2017," National Oceanic and Atmospheric Administration (NOAA) Technical Memorandum, GLERL-170, December 12, 2017).

[8] Wang et al., 2017.

municipal, agricultural and industrial activities, most of which is treated and returned.[9] The U.S. Great Lakes maritime economy supports 311,000 jobs spanning every sector, from tourism and recreation to marine transportation. This translates to $8.8 billion in wages.[10]

Understanding how the water flows through this network of fresh water bodies further underscores the uniqueness of the region. The lakes are a series of steps leading from Lake Superior, at 601 feet above sea level, down to the Atlantic Ocean, as shown in figure 1.2. The St. Marys River flows from Lake Superior to Lake Huron. Lakes Michigan and Huron are connected by the Straits of Mackinac with levels that rise and fall together. The St. Clair and Detroit rivers, with Lake St. Clair in between, connect Lake Huron with Lake Erie. The Niagara River then links Lake Erie with Lake Ontario, including Niagara Falls. From Lake Ontario, water flows through several smaller lakes and into the St. Lawrence River, which converges with the Ottawa River and flows to the Atlantic Ocean in the Gulf of St. Lawrence. Monthly water levels within the lakes can vary by four to six feet, according to 104 years of record-keeping.[11]

Figure 1.2. Great Lakes System Profile

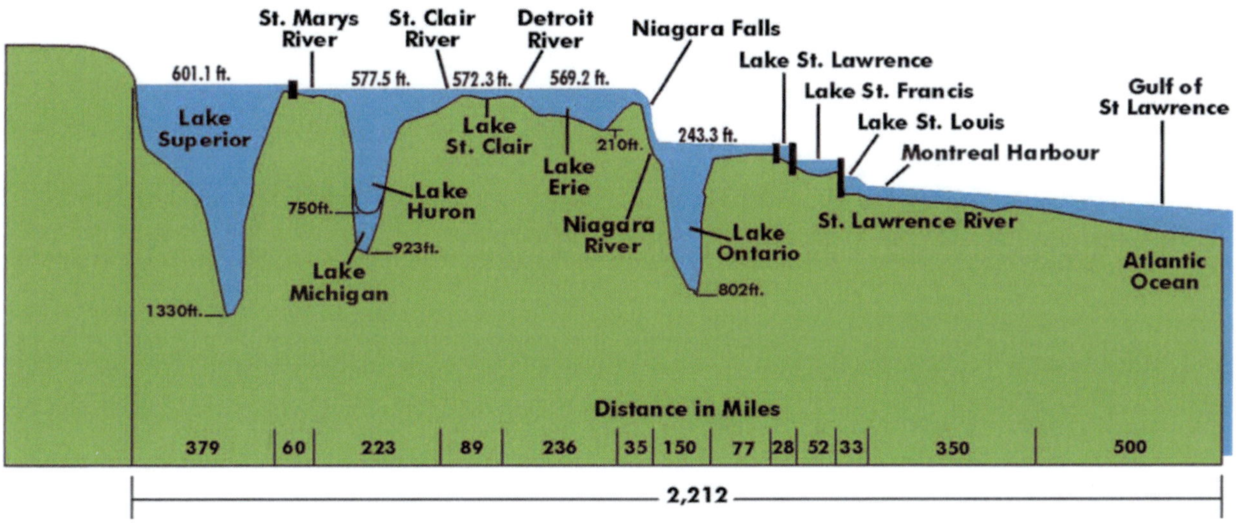

SOURCE: Reproduced from Keith Kompoltowicz, "Understanding Great Lakes Water Level Fluctuations and the Need for Coastal Resiliency," *Proceedings*, Vol. 79, No. 3, Winter 2022.

---

[9] Terry Gibb, "Lakes Appreciation Month: The Great Lakes Facts and Features," Michigan State University Extension, July 20, 2015.

[10] NOAA Office for Coastal Management, "Great Lakes," 2023; NOAA, Office for Coastal Management, *2022 Marine Economy Report*, undated.

[11] For the purposes of this analysis, the geographic scope of the project includes the Great Lakes, connecting lakes (e.g., Lake St. Claire), and rivers that flow directly into the lakes (e.g., Detroit River and St. Marys River) or out of them (St. Lawrence Seaway).

The Great Lakes is essentially a closed system given that less than 1 percent of the total water volume outflows in a year. With annual replacement rates of less than 1 percent, pollutants introduced into the Great Lakes can, over time, concentrate, resuspend, and infiltrate the biologic system through annual lake turnover, navigational dredging, and cycling through the food web. Past spills of oil and hazardous substances resulted in highly elevated levels of such byproducts as polycyclic aromatic hydrocarbons, heavy metals (i.e., mercury), and toxins (including polychlorinated biphenyls).[12]

Figure 1.3. Number of Years that Pollutants Could Be Retained in Each of the Great Lakes

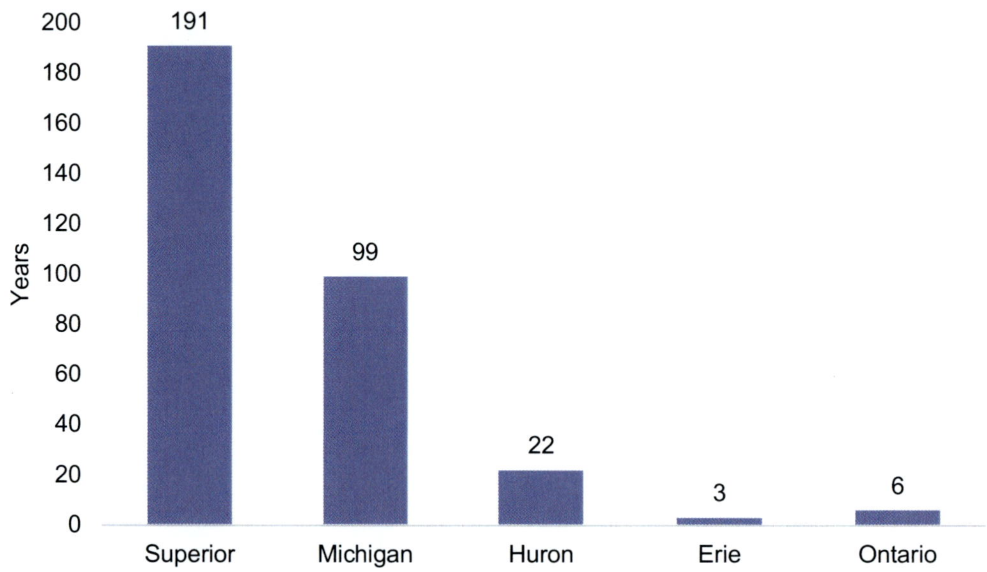

SOURCE: Features information from NOAA, Great Lakes Environmental Research Laboratory, "About Our Great Lakes, Lake by Lake Profiles," June 2004.
NOTE: This figure indicates each lake's replacement time in years. *Replacement time* is the amount of time it takes for lakes to get rid of pollutants.

Climate change is expected to affect the shipping industry and the infrastructure that the industry relies on in several ways. Although the effects of climate change are beyond the scope of this study, we note that climate change will affect the risk of oil spills and the response capabilities needed. Historically, the water levels in the Great Lakes fluctuate with precipitation levels, snow runoff, drought, withdrawals, and evaporation rates (Figure 1.4).[13]

---

[12] Thomas Rayburn, "Integrating Data, Resources, and Capabilities in the Great Lakes: Impacting Federal, State, Tribal, Local, and Private Partnerships Through the Area Committee," Great Lakes Commission, 2003.

[13] NOAA, Great Lakes Environmental Research Laboratory, "Water Levels," brochure, undated c; EPA, "Climate Change Indicators: Great Lakes Water Levels and Temperatures," webpage, August 2, 2022b.

Figure 1.4. Historical Water Levels in the Great Lakes

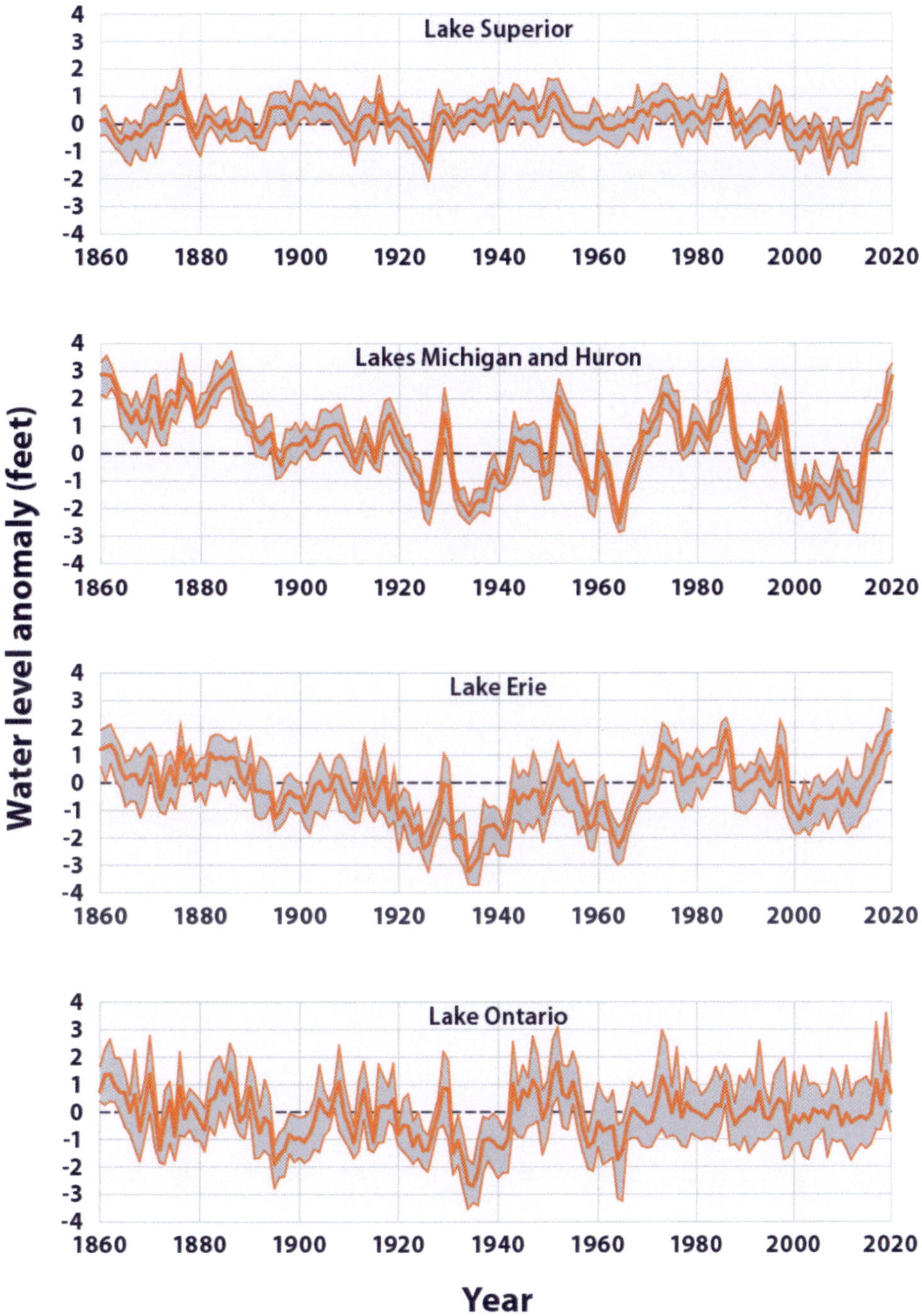

SOURCE: Reproduced from EPA, 2022b.

Climate change is anticipated to increase the lake level extremes, and shifts between extremes will occur more rapidly.[14] These rapidly fluctuating water levels will affect shipping, navigation, fisheries, tourism, recreation, and such coastal infrastructure as roads, piers, and wetlands in the region. For example, during periods of low water levels, large ships might not traverse some areas in the region.[15] Additionally, if water levels significantly drop, or become more variable, this could contribute to an increase in the frequency of vessel groundings and associated incidents, to potentially include oil spills. Similarly, less-predictable ice conditions could also contribute to vessel casualties. Although the effects on oil spills and emergency response is unclear, it is estimated that for each 1-inch decrease in water levels, cargo ship capacity is reduced by several hundred tons, resulting in thousands of dollars in lost profits per ship.[16] Low water levels will also effect marinas and canals and might require dredging to sustain shipping and recreational boating.[17]

Climate change also will affect the extent and duration of ice cover on the Lakes. Ice cover has decreased in the last 30 years with substantial year-to-year variability. Climate change will likely shorten the ice season, although there are several factors that will continue to affect year-to-year variability. Warmer air temperatures suggest that there might be less ice on average, but the occurrence of such factors as Arctic air blasts, El Nino-Southern Oscillation, and Arctic Oscillation, which are difficult to predict, can still lead to extensive ice coverage in any given year. A shorter ice season will likely extend the shipping season, but shippers will have to remain adaptable because ice cover will continue to vary year-to-year. Reduced ice during winter seasons might also result in shorelines that are exposed to winter winds and storms, increasing erosion that might threaten shoreline infrastructure.[18]

Great Lakes Sensitivities

According to Great Lakes subject-matter experts (SMEs), the enormous trade, industrialization, and transportation infrastructure growth that the region has experienced over

---

[14] Kim Channell, John DelPizzo, Laura Briley, Richard Rood, Jenna Jorns, and Kate Hutchens, "Lake Ontario: A Summary of Anticipated Future Climate Conditions," Great Lakes Integrated Sciences and Assessments, September 2022.

[15] J. Wang, X. Bai, H. Hu, A. Clites, M. Colton, and B. Lofgren, "Temporal And Spatial Variability of Great Lakes Ice Cover, 1973–2010," *Journal Of Climate,* Vol. 25, No. 4, 2012.

[16] D. Marchand, M. Sanderson, D. Howe, and C. Alpaugh, "Climatic Change and Great Lakes Levels The Impact On Shipping," *Climatic Change*, Vol. 12, No. 2, 1988, in Channell et al., 2022.

[17] S. Changnon, "Changes in Climate and Levels of Lake Michigan: Shoreline Impacts at Chicago," *Climatic Change,* Vol. 23, No. 3, 1993; A. Bartolai, L. He, A. Hurst, L. Mortsch, R. Paehlke, and D. Scavia, "Climate Change as a Driver of Change in the Great Lakes St. Lawrence River Basin," *Journal Of Great Lakes Research*, 2015; in Channell et al., 2022; and Dawn Nelson, Heather Elmer, Rebecca Held, Danielle Forsythe, and Shauna Casey, "Laurentian Great Lakes Basin Climate Change Adaptation," NOAA Technical Memorandum GLERL-153, March 2011.

[18] Channell et al., 2022.

the past century was lacking in environmental awareness and resulted in large-scale pollution and contamination. For example, in the mid-1900s, Cleveland's Cuyahoga River reported nine fires, heavy metals entering the lakes, and large amounts of phosphorous that caused eutrophication.[19] After decades of public outcry and several initiatives, including the Clean Water Act, Federal Water Pollution Control Act, and Great Lakes Quality Agreements of the late 1970s, water quality has improved.

Because of the ongoing elevated environmental sensitivity of the region, the spill category classifications in the National Oil and Hazardous Substances Pollution Contingency Plan (NCP) (40 C.F.R. 300) have been reduced from salt water environment thresholds. For example, a major spill is classified as more than 10,000 gallons, which is an order of magnitude less than the salt water coastal zone major threshold of 100,000 gallons.

More recently, oil production has increased as a result of the extraction of shale and oil sand substances from regional deposits. Much of the crude oil products are transported through the Great Lakes region via several modes but primarily through pipelines and rail cars. These modes cross the Great Lakes or their tributaries and run alongside or near the Great Lakes' shorelines. A 2010 pipeline spill that released more than 800,000 gallons of heavy oil was arguably a sentinel event for the region, precipitating a renewed call for action and resulting in legislative action.[20]

The International Joint Commission's Great Lakes Water Quality Board completed its third Great Lakes Regional Poll in 2021. The 2021 poll results provide a snapshot of Great Lakes residents' views on topics including the importance of protecting environmental health and the need to protect the Lakes' water quality for leisure and recreation, fish, and wildlife, and the economy. The board polled 4,550 Canadian and U.S. residents in the Great Lakes region. Figure 1.5 provides a summary of the key findings, which underscores that a majority of respondents seek more protections for the Great Lakes.[21]

---

[19] *Eutrophication* occurs when the environment becomes enriched with nutrients, increasing the amount of plant and algae growth, according to NOAA, which can starve animal life of oxygen (NOAA, National Ocean Service, "What is Eutrophication?" webpage, undated a).

[20] Jerome A. Popiel, "Great Lakes Environmental Response," *Proceedings*, Vol. 79, No. 3, Winter 2022.

[21] Canada and the United States created the International Joint Commission (IJC) to manage lake and river systems along the border. The IJC is guided by the Boundary Waters Treaty, signed by Canada and the United States in 1909, which provides general principles for preventing and resolving transboundary issues. The IJC has two main responsibilities: approving projects that affect water levels and flows across the boundary, and investigating transboundary issues and recommending solutions. The IJC takes into account the needs of a wide variety of water uses, including drinking water, commercial shipping, hydroelectric power generation, agriculture, ecosystem health, industry, fishing, recreational boating, and shoreline property (International Joint Commission, "Role of the IJC," webpage, undated). International Joint Commission, Great Lakes Water Quality Board, 2021 Great Lakes Regional Poll, November 2021a; International Joint Commission, Great Lakes Water Quality Board, *2021 Great Lakes Regional Poll Online Poll Final Report*, November 2021b.

Figure 1.5. Key Findings from the 2021 Great Lakes Regional Poll

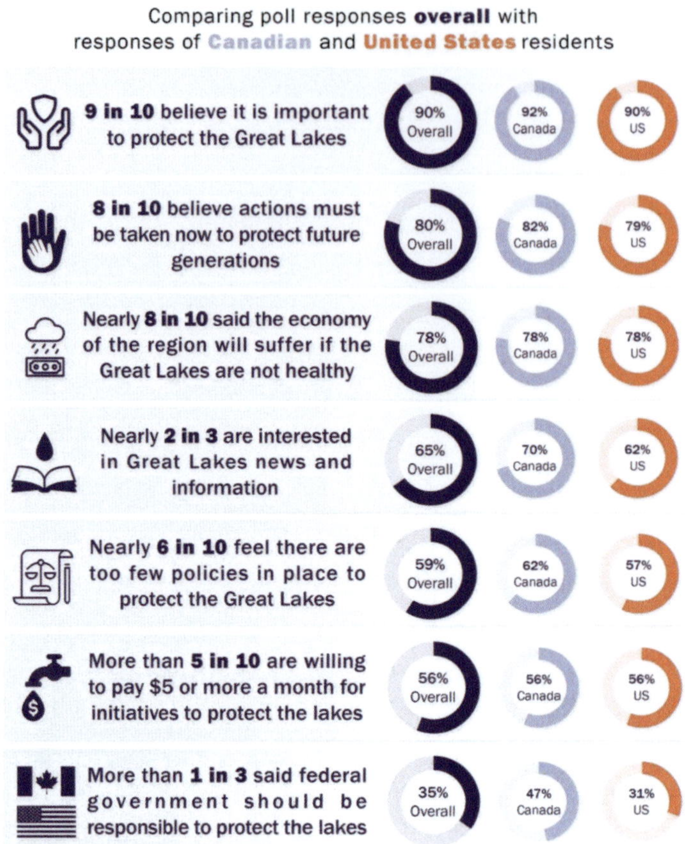

SOURCE: Reproduced from International Joint Commission, Great Lakes Water Quality Board, 2021 Great Lakes Regional Poll, Key Findings, November 2021c.

Project Objective and Approach

USCG Headquarters, Office of Marine Environmental Response Policy asked HSOAC, a federally funded research and development center for the U.S. Department of Homeland Security (DHS) operated by the RAND Corporation, to gather and assess information on the risks of oil spills in the Great Lakes region and response capabilities.

The project objective was to provide information and analysis that helps the USCG understand the scope of oil spill risks in the region; the roles, capabilities, and capacities of oil-spill response partners; and gaps and mitigation strategies around oil-spill response in fresh water and ice.

Approach and Methodology

To identify capabilities, gaps, and mitigation strategies, we applied the following definitions:[22]

- A *capability* is defined as "the means to accomplish a mission, function, or objective."
- A *capability gap* occurs when "a capability that is required by DHS and/or its stakeholders or partners to perform a mission, function or objective" does not currently exist and there are no plans to provide it in existing programs.

To characterize overall capabilities and capacity to respond to spills in the Great Lakes, and to assess gaps and ways to mitigate them, we conducted several tasks. To inform these tasks, HSOAC researchers performed a document and literature review and structured interviews (Table 1.1) with Canadian and U.S. federal agency personnel and other stakeholders in industry and communities in the Great Lakes region.[23]

Table 1.1. Stakeholder Groups Interviewed

Stakeholder	Interview Status
USCG Headquarters Office (Office of Marine Environmental Response Policy)	Interviewed
USCG Ninth District Response Advisory Team	Interviewed
USCG Ninth District Public Affairs Office	Interviewed
USCG Marine Environmental Response	Interviewed
USCG Sector Lake Michigan	Interviewed
USCG Sector Detroit	Interviewed
USCG Sector Sault Sainte Marie	Interviewed
USCG Sector Buffalo	Interviewed
Lake Carriers' Association	Interviewed
NOAA	Interviewed
Marine Pollution Control (based in Detroit)	Interviewed
Southeast Michigan Council of Governments	Interviewed
Environment and Climate Change Canada	Interviewed
Algoma Central Corporation	Interviewed

---

[22] For reference, a *capability gap* occurs when "a capability that is required by DHS and/or its stakeholders or partners to perform a mission, function or objective" does not currently exist and there are no plans to provide it in existing programs (DHS, *Department of Homeland Security Manual for the Operation of the Joint Requirements Integration and Management System*, DHS Instruction Manual 107-01-001-01, April 4, 2016, p. 3).

[23] Relevant and current oil-spill response capability and capacity documentation resides almost exclusively in the various governmental area response plans, oil-spill response organization operational planning and guidance documentation, government sponsored websites, and governmental regulation, policy, and TTP guidance. Therefore, we were limited to what was provided, which might be regarded as a limitation to this analysis. Field visits and regional workshops were beyond the scope of this study.

Stakeholder	Interview Status
IISD Experimental Lakes Area	Interviewed
Canadian Environmental Response	Interviewed
Eastern Canada Response Corporation (ECRC)	Interviewed
Chamber of Marine Commerce (binational, based in Canada)	Interviewed
Ontario Ministry of Environment, Conservation and Parks	Contacted but did not interview
EPA Region 5	Contacted but did not interview
Alliance for the Great Lakes	Contacted but did not interview
Enbridge, Inc.	Contacted but did not interview

Task 1. Describe the Scale and Scope of the Problem

We drew on a combination of documents and interviews to characterize the flow of oil in the Great Lakes and their immediate watershed; roughly what quantities of oil are moved by pipeline, by vessel, and by other means in and around the Great Lakes; and any indications of potential large-scale changes in the next couple of decades. Documents were provided or obtained directly from the sponsor, government sponsored websites, SMEs, oil pollution response organizations, and relevant publicly available research. Citations and references are provided throughout the report and listed at the end of the report. The majority of material resides with government institutions and organizations that are primarily responsible for oil-spill response operations, regulation, and policy, to include USCG, NOAA, EPA, and cognizant Canadian governmental organizations (see Table 1.1).

Task 2. Describe the Oil-Spill Response Capabilities and Capacities of Government Agencies and Private Companies on Both Sides of the U.S.-Canadian Border

This analysis involved reaching out to stakeholders in various agencies and companies to characterize their capabilities, in addition to drawing on any available documentation. Different types of cleanup capabilities were considered, given that capabilities and capacities might be affected by specific environmental, seasonal, and other factors and because the field of fresh water oil-spill response science is less mature than that of salt water. The analysis also took into account the physical locations of capabilities and the speed with which they can be deployed where they are needed. The ability of deploying spill response from outside the region is also considered, along with the timelines required for such assistance to reach the Great Lakes and the extent to which outside entities can collaborate effectively. Although the USCG has a clear oil-spill response chain of command, it is less clear how well coordination across contractors, responsible parties, states, and across the Canada–U.S. border might function. The effectiveness of their collaboration, including across an international border, can shape the extent to which their collective capacities might be greater than or less than the sum of their individual capacities.

Task 3. Develop and Analyze a Series of Scenarios to Assess Gaps in Response Capabilities and Capacity

To assess gaps in response capabilities and capacities, we developed and analyzed a series of scenarios that have been used to practice and assess oil-spill response in the Great Lakes. A scenario framework was built for a more multidimensional approach to exercising response capabilities. Ideally, these dimensions can be developed with defined levers that can be adjusted depending on regional focus, priority, and stakeholder participation.

The dimensions are

- volume of spill
- location of spill (to determine Canadian engagement)
- type of product
- presumed flow or fate of pollutant.
- weather conditions (to include ice, wind, waves)
- location sensitivities (e.g., environmental, ecological, drinking water intake)
- wildcards (e.g., terrorism, holiday weekend, road transit impeded, marine spill already underway, change in traffic patterns because of climate change or energy demand, other mission surge event).

The dimensions can then be used to create exercise scenarios that can be tailored to address regional concerns, changing technologies, and partner input. Finally, the framework can be used to develop a case study scenario as an exemplar of how to use the framework to build out more-robust scenarios for future exercises.

Task 4. Analyze How Gaps Can Be Mitigated

This task entailed identifying emerging technologies or tactical approaches, which required additional research to evaluate their efficacy. Approaches included changes to statutory authorities, improved training, intensified coordination among multiple U.S. and Canadian agencies and the private sector, fresh water-specific oil-spill scientific research, accelerated deployment timelines, enhanced characterization of oil distribution over time through the use of additional sensors and spill modeling, and facilitating the movement of assets from outside the Great Lakes. We also considered mitigations from the literature on fresh water or ice oil-spill response plans, technology, and resources from non-North American nations.

The Organization of This Report

The remainder of this report is organized as follows: Chapter 2 describes the scale and scope of the challenge for oil-spill response. Chapter 3 summarizes the capabilities and capacity of government agencies and other responsible parties on both sides of the U.S.-Canadian border. Chapter 4 is an analysis of whether the responsible parties that do much of the response work can get equipment to potential spill sites within the required response time. Chapter 5 walks through how to build practice scenarios for exercises that could identify any gaps in capacities and capabilities for oil-spill response. Chapter 6 describes potential challenges for stakeholders. Chapter 7 provides some closing considerations. Appendix A gives more detail on the capabilities of oil-spill responders. Appendix B provides instructions and tools for developing scenarios that are used to practice and improve organizational preparedness for responding to oil spills. Finally, Appendix C contains the interview protocol used during the structured discussions with SMEs.

Chapter 2. Describing the Scale and Scope of the Challenge

This chapter examines the scale and scope of the challenge of responding to oil spills in the Great Lakes region. We first discuss the definition of *oil* as it applies to oil-spill response. Then, we characterize the locations and flows of oil throughout the Great Lakes region. Finally, we assess likely sources of oil spill risk and discuss future changes to oil flow throughout the region.

Oil Definitions

The type of oil involved in a spill is important because the type of response required will depend heavily on the product. The term *oil* can refer to a broad set of substances. We examined scientific definitions of oil, and U.S. and Canadian regulatory definitions. By examining the definitions, we aimed to better understand the circumstances under which U.S. and Canadian authorities require a response and identify any discrepancies in how oil is defined.

Several key U.S. regulations govern response to oil spills, including the Oil Pollution Prevention Regulation, the Clean Water Act, and the Oil Pollution Act of 1990, which amended the Clean Water Act.[24] Notably, the Clean Water Act established broad reporting requirements that extend beyond oil to cover a wide variety of discharged substances, as follows:[25]

- discharges that cause a sheen or discoloration on the surface of a body of water
- discharges that violate applicable water quality standards
- discharges that cause a sludge or emulsion to be deposited beneath the surface of the water or on adjoining shorelines.

U.S. pollution laws distinguish between oil spills, covered by the above regulations, and hazardous substance releases, which are covered by the Comprehensive Environmental Response, Compensation, and Liability Act (CERCLA). Chemicals covered by CERCLA, such as benzene, are components of crude oil. In some cases, such as the response to the pollution threat posed by the sunken tank barge Argo, both sets of regulations apply, which can complicate the response.[26]

Relevant Canadian regulations included the Canada Oil and Gas Operations Act of 1985, the Canada Shipping Act of 2001, and the Marine Liability Act of 2001. The definitions of oil from each of these regulations are presented in Table 2.1.

---

[24] USCG, National Pollution Funds Center, "Oil Pollution Act of 1990 (OPA)," webpage, undated.

[25] 33 U.S.C. 1320, Federal Water Pollution Control Act, Oil and Hazardous Substance Liability, Section 311 (1), 2002; EPA, "Discharge of Oil Regulation Overview," webpage, Environmental Protection Agency Archive, undated.

[26] USCG, Argo After-Action Report, undated a, Not available to the general public, p. 2.

Table 2.1. Regulatory Definitions of Oil

Source Type	Source	Definition of Oil
U.S. Regulatory Definitions	Oil Pollution Prevention Regulation	"oil of any kind or in any form, including, but not limited to: fats, oils, or greases of animal, fish, or marine mammal origin; vegetable oils, including oils from seeds, nuts, fruits, or kernels; and, other oils and greases, including petroleum, fuel oil, sludge, synthetic oils, mineral oils, oil refuse, or oil mixed with wastes other than dredged spoil."[a]
	Federal Water Pollution Control Act (Clean Water Act)	"oil of any kind or in any form, including, but not limited to, petroleum, fuel oil, sludge, oil refuse, and oil mixed with wastes other than dredged spoil."[b]
	Oil Pollution Act of 1990	"oil of any kind or in any form, including petroleum, fuel oil, sludge, oil refuse, and oil mixed with wastes other than dredged spoil, but does not include any substance which is specifically listed or designated as a hazardous substance under subparagraphs (A) through (F) of section 101(14) of the Comprehensive Environmental Response, Compensation, and Liability Act (42 U.S.C. 9601) and which is subject to the provisions of that Act [42 U.S.C. 9601 et seq.]."[c]
Canadian Regulatory Definitions	Canada Shipping Act, 2001	"petroleum in any form, including crude oil, fuel oil, sludge, oil refuse and refined products."[d]
	Marine Liability Act, 2001	"oil of any kind or in any form and includes petroleum, fuel oil, sludge, oil refuse and oil mixed with wastes but does not include dredged spoil"[e]
	Canada Oil and Gas Operations Act, 1985	"(a) crude petroleum regardless of gravity produced at a well-head in liquid form, and (b) any other hydrocarbons, except coal and gas, including hydrocarbons that may be extracted or recovered from surface or subsurface deposits, including deposits of oil sand, bitumen, bituminous sand, oil shale and other types of deposits"[f]

[a] 33 U.S.C. § 1251, Title 40, Chapter I, Subchapter D, Part 112, Oil Pollution Prevention, as amended 2002.
[b] 33 U.S.C. § 1320, Federal Water Pollution Control Act, Section 311 (1), Oil and Hazardous Substance Liability, 2002.
[c] 33 U.S.C. § 2701, Chapter 40, Oil Pollution, Definitions.
[d] Statutes of Canada, Canada Shipping Act, Chapter 26, Part 8, Pollution Prevention and Response—Department of Transport and Department of Fisheries and Oceans, 2001.
[e] Statutes of Canada, Marine Liability Act, Chapter 6, 2001.
[f] Revised Statutes of Canada, Canada Oil and Gas Operations Act, Chapter O-7, 1985.

Notably, U.S. definitions are fairly broad and tend to include both petroleum and non-petroleum-based products, while both the Canada Shipping Act and the Canada Oil and Gas Operations Act of 1985 focus narrowly on petroleum and petroleum products. For example, an EPA article on types of oil relevant to an oil-spill response focuses on "crude oils and refined petroleum products" relevant to a release (e.g., spill into a body of water).[27] The National Aeronautics and Space Administration (NASA) Earth Science and Applied Sciences defines an oil spill as "the release of a liquid petroleum hydrocarbon into the environment, especially

[27] EPA, "Types of Oil," webpage, January 24, 2022.

marine areas, because of human activity.[28] The term is usually applied to marine oil spills, where oil is released into the ocean or coastal waters, but spills may also occur on land."[29] As another example, a NOAA article discusses the general chemical composition of oil relevant to an oil spill

> Crude oils, while mixtures of thousands of types of molecular compounds, are predominantly composed of only two types of atoms: hydrogen (H) and carbon (C). Molecular compounds composed exclusively of these two elements are called hydrocarbons. Petroleum hydrocarbons are predominantly one of two types, aromatics or alkanes.[30]

We also reviewed exercise and response after-action reports from U.S. and Canadian sources and found that these focused on petroleum-based oils. Although the USCG's 2015 Great Lakes Area Contingency Plan does include nonpetroleum oils, it places more emphasis on petroleum oil.[31] Overall, we assess that the primary risk from an oil spill will come from a spill of petroleum oil, so we have focused our analysis on petroleum and petroleum products.

The USCG's Great Lakes Area Contingency Plan classifies types of petroleum oil as persistent or nonpersistent based on distillation criteria; it further categorizes persistent oils based on specific gravity.[32] Table 2.2 defines these categories of oil.

Table 2.2. Types of Petroleum Oil

Oil Type	Definition
Nonpersistent/ Group I	A petroleum-based oil such that at least 50 percent by volume distills at a temperature of 340°C and 645°F, and at least 95 percent by volume distills at a temperature of 370°C and 700°F.
Persistent	A petroleum-based oil that does not meet the distillation criteria for nonpersistent oils. Persistent oils are further categorized as follows: • Group II: specific gravity of less than 0.85. Examples: gasoline, kerosene, and Nigerian Light Crude. • Group III: specific gravity of at least 0.85 and less than 0.95. Examples: Arabian and Kuwait crude oil.

[28] The Disasters program area of NASA's Earth Science Applied Sciences Program uses Earth-observing data and applied research to improve the prediction of, preparation for, response to and recovery from hazards and disasters around the world. Before, during and after disasters strike, NASA coordinates with decisionmakers and local governments, providing actionable data to recover from disaster impacts and build resilient communities. A form of pollution, oil spills might be caused by releases of crude oil from tankers, offshore platforms, drilling rigs or wells. Spilled substances might be refined petroleum products, such as gasoline and diesel fuel, and their by-products—heavier fuels used by large ships such as bunker fuel or oily refuse of any kind (NASA, Earth Science Applied Sciences, "Supporting Recovery Efforts After Oil Spills," webpage, undated).

[29] NASA, undated.

[30] NOAA, Office of Response and Restoration, "Chemistry of an Oil Spill," webpage, August 21, 2012.

[31] USCG, Great Lakes Area Contingency Plan, April 13, 2015, pp. 6, 91.

[32] USCG, 2015, pp. 6, 91. *Specific gravity* is defined as the ratio of the weight of a given volume of oil to the weight of the same volume of water at a given temperature. The density of fuel, relative to water, is called specific gravity. The specific gravity of water is defined as 1. Because specific gravity is a ratio, it has no units.

Oil Type	Definition
	• Group IV: specific gravity of at least 0.95 and less than or equal to 1.0. Examples: Bunker C and #6 fuel oil. • Group V: specific gravity greater than 1.0. Examples: Carbon Black and asphalt products.

SOURCE: Adapted from USCG, 2015, pp. 6, 91.

Oil Locations and Movements Through the Great Lakes Region

The Great Lakes region is both a source of oil and a transportation route for a great quantity of oil. The region contains numerous oil wells and petroleum refineries.[33] *Tight oil plays*—low-permeability rocks from which oil can be extracted via directional drilling and hydraulic fracturing—are also present.[34] The U.S. locations of these sources, shown in Figure 2.1, are concentrated between Lakes Michigan and Huron and south of Lake Erie. There are few, if any, noted sources that might affect Lakes Superior or Ontario.

Figure 2.1. U.S. Oil Plays, Wells, and Refineries in the Great Lakes Region

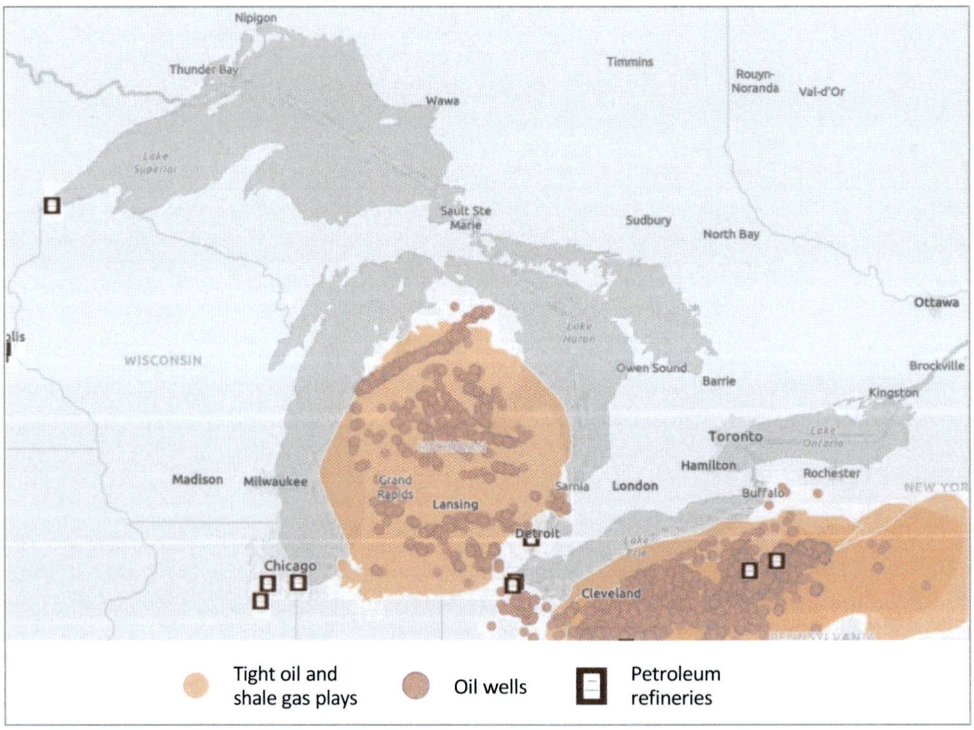

---

[33] Numerous oil wells have been abandoned and need to be plugged, posing additional potential risk; for instance, approximately 19,000 abandoned oil wells have been identified in Ohio, and it is believed that more likely exist (Ohio Department of Natural Resources, "Orphan Well Program," webpage, undated).

[34] Jeremy Boak and Robert Kleinberg, "4–Shale Gas, Tight Oil, Shale Oil and Hydraulic Fracturing," *Future Energy (Third Edition)*, 2020.

SOURCE: Created with U.S. Energy Information Administration, "All Energy Infrastructure and Resources," web mapping application, updated April 20, 2023.

In addition, oil is moved through the Great Lakes region using several means, including pipelines, vessels, rail, and road. The following sections describe each of these means of transport in detail.

Pipelines

Of the four means of oil transport examined, pipelines are the primary means of transport in the Great Lakes region. Several major pipelines provide an estimated 90 percent of crude oil volume to U.S. refineries in the Great Lakes region.[35] Although pipelines are generally considered the safest means of oil transport, the large volume of oil they carry also means they pose the greatest oil spill risk in the Great Lakes.[36] Note that data on the throughput of pipelines within the region is generally proprietary and was not available for all pipelines.

Crude oil and petroleum products are carried via different pipelines to prevent impurities in crude oil from contaminating petroleum products, the composition of which is tightly regulated. However, different types of crude oil can be carried through the same pipeline.[37] Figure 2.2 shows the pathways of pipelines carrying crude oil and petroleum products through the region. One of the largest pipeline systems is the Enbridge Mainline system, which carries crude oil from Edmonton, Alberta, to locations throughout Ontario, Quebec, and the U.S. Midwest. It has the capacity to transport 2.85 million barrels per day.[38] Other crude oil pipelines in the area are the MarkWest Michigan Crude Pipeline, which connects to Enbridge's Line 5 pipeline (part of the Mainline system) within Michigan; BP's Chicap Pipeline, which moves oil within Illinois; and Sunoco's Mid Valley Pipeline, which moves crude oil between Texas and Michigan.[39] Refined petroleum products are also transported throughout the region via pipelines operated by several companies, including BP, Buckeye Partners, Enterprise Products, Explorer Pipeline, Koch Pipeline, Marathon, Midstream Partners, Sunoco, West Shore Pipeline Company, and Wolverine Pipeline Company. Pipeline locations point to few potential sources that could affect Lake Superior and a large portion of Lake Huron.

---

[35] Hazardous Material Transport Outreach Network (HAZMATON), "Crude Oil Transport by Pipeline in the Great Lakes," webpage, undated.

[36] HAZMATON, undated.

[37] Bradley Hull III, *Status of Infrastructure Related to Crude Oil Transportation in the Great Lakes/St. Lawrence River Region*, May 31, 2017.

[38] Enbridge, *The Mainline Pipeline System*, fact sheet, undated c.

[39] Tip of the Mitt Watershed Council, "Enbridge Energy Line 5," undated; BP, "Our Pipelines," webpage, undated; U.S. Energy Information Administration, 2023; Enbridge, *Enbridge's Energy Infrastructure Assets*, April 19, 2023.

Figure 2.2. Locations of Crude Oil and Petroleum Product Pipelines in the Great Lakes Region

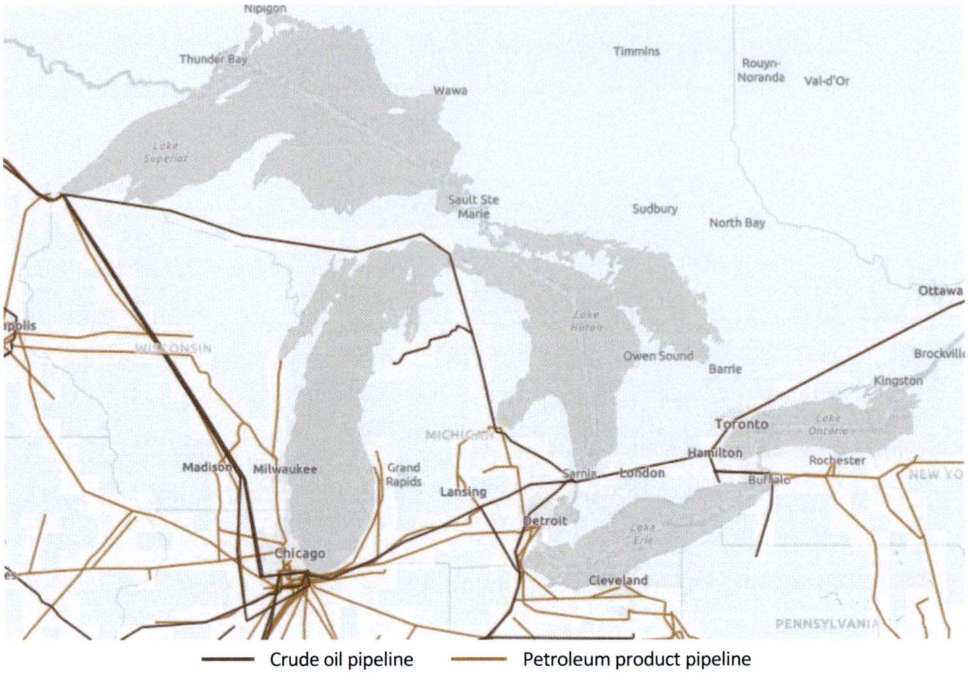

SOURCE: Created with U.S. Energy Information Administration, 2023.

Vessels

Crude oil is not shipped on the Great Lakes, but a variety of petroleum products are shipped to or from Great Lakes ports.[40] Table 2.3 shows a summary of the types of petroleum products carried as freight in 2020 and whether the traffic was considered foreign or domestic. Foreign traffic comprises inbound traffic from foreign sources and outbound traffic from the U.S. to foreign countries.[41] Domestic freight traffic is broken down into the following categories:[42]

- coastwise: domestic traffic that transits via an ocean or the Gulf of Mexico. This can include traffic between Great Lakes ports and seacoast ports.
- lakewise: traffic between U.S. ports on the Great Lakes
- internal: traffic which take place solely on inland waterways
- intraport: traffic within the confines of a port.

Overall, most of the freight carried in 2020 was domestic lakewise vessel traffic.

---

[40] John Frittelli, *Shipping on the Great Lakes and St. Lawrence Seaway: An Update*, Congressional Research Service, R47550, May 9, 2023; U.S. Army Corps of Engineers, Waterborne Commerce of the United States, 2020, Part 3–Waterways and Harbors: Great Lakes, 2022, p. xv.

[41] U.S. Army Corps of Engineers, 2022, p. xv.

[42] U.S. Army Corps of Engineers, 2022, p. xv.

In addition, Figure 2.3. shows tanker tracks through U.S. territory in the Great Lakes region in 2018. This analysis was based on vessel traffic data using the vessels' automated identification system (AIS) from the Bureau of Ocean Energy Management (BOEM) and NOAA.[43]

---

[43] AIS is a digital positional awareness system operating in the Very High Frequency maritime band. Its purpose is to help identify ships, assist in target tracking and search and rescue operations, simplify information exchange, and provide situational awareness *(International Maritime Organization (IMO), A 29/Res.1106)*. It was first developed as a collision avoidance tool to enable commercial vessels to see each other more clearly in all conditions. This system, combined with a shore station, also enables the management of manage maritime traffic and reduce the hazards of marine navigation (North Atlantic Treaty Organization, "AIS (Automatic Identification System) overview," webpage, 2021).

Table 2.3. Freight Traffic in 2020, Thousands of Barrels

Commodity	Grand Total	Foreign: Inbound: Canadian	Foreign: Inbound: Overseas	Foreign: Outbound: Canadian	Foreign: Outbound: Overseas	Domestic: Coastwise	Domestic: Lakewise	Domestic: Internal	Domestic: Intraport
Total petroleum and petroleum products	20,111	2,465	86	1,596	1,121	120	12,693	709	1,321*
Subtotal petroleum products	20,111	2,465	86	1,596	1,121	120	12,693	709	1,321*
Gasoline	1,939	80		143			1,590		126
Distillate fuel oil	3,089	498		486			566	601	938
Residual fuel oil	531	280				120	17		114
Lube oil and greases	46							23	23
Naphtha and solvents	132			132					
Asphalt, tar, and pitch	9,225	120					8,917	74	114
Petroleum coke	5,050	1,487		835	1,121		1,596	11	
Hydrocarbon and petrol gases, liquefied and gaseous	92		86				6		

SOURCE: U.S. Army Corps of Engineers, 2022, p. xv.
NOTE: The sum of domestic intraport petroleum products listed sums to 1,315; unable to determine the source of the discrepancy between this and the total/subtotal provided.

Figure 2.3. Oil Tanker AIS Tracks in the Great Lakes Region, 2018

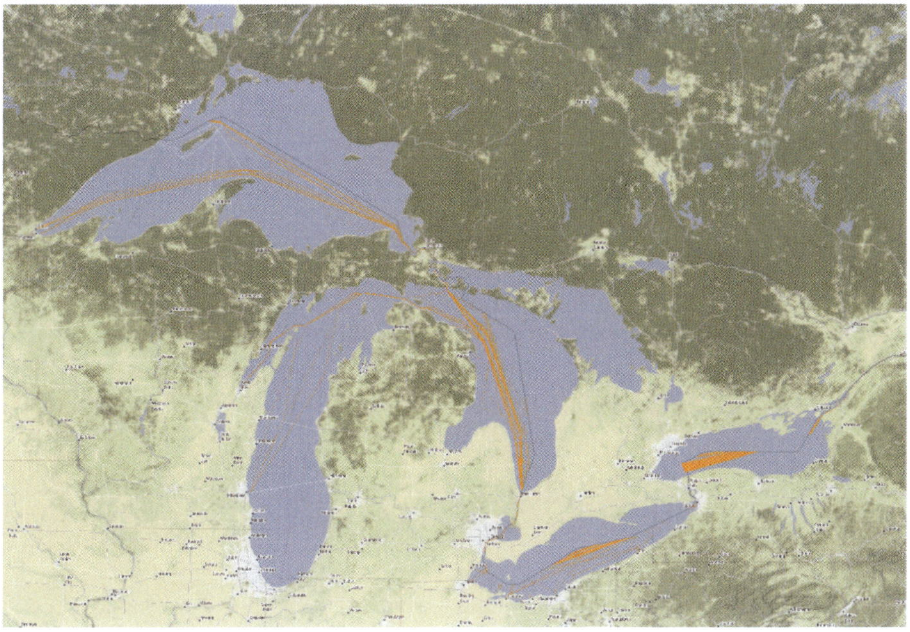

SOURCE: Features information from NOAA and BOEM, "AISVesselTracks2018GreatLakes," database, undated.

Rail

Two crude oil rail terminals exist within the Great Lakes region on the U.S. side: the Channahon Terminal in Lorenzo, Illinois, and the Toledo Refinery in Toledo, Ohio.[44] Figure 2.4 shows the locations of these rail terminals along with the U.S. rail network.

Several types of rail cars are in use in the United States, including the DOT-117 and DOT-117R, DOT-111, and CPC-1232.[45] Rail car capacities vary, but they generally can hold approximately 680–715 barrels of oil.[46]

---

[44] U.S. Energy Information Administration, 2023.

[45] U.S. Department of Transportation, Bureau of Transportation Statistics, *Fleet Composition of Rail Tank Cars Carrying Flammable Liquids: 2022 Report*, 2022.

[46] The Greenbrier Companies, *28,600 Gallon DOT 117 Tank Car*, fact sheet, January 2021; Congressional Research Service, *U.S. Rail Transportation of Crude Oil: Background and Issues for Congress*, R43390, December 4, 2014.

Figure 2.4. U.S. Crude Oil Rail Terminals and Railroads

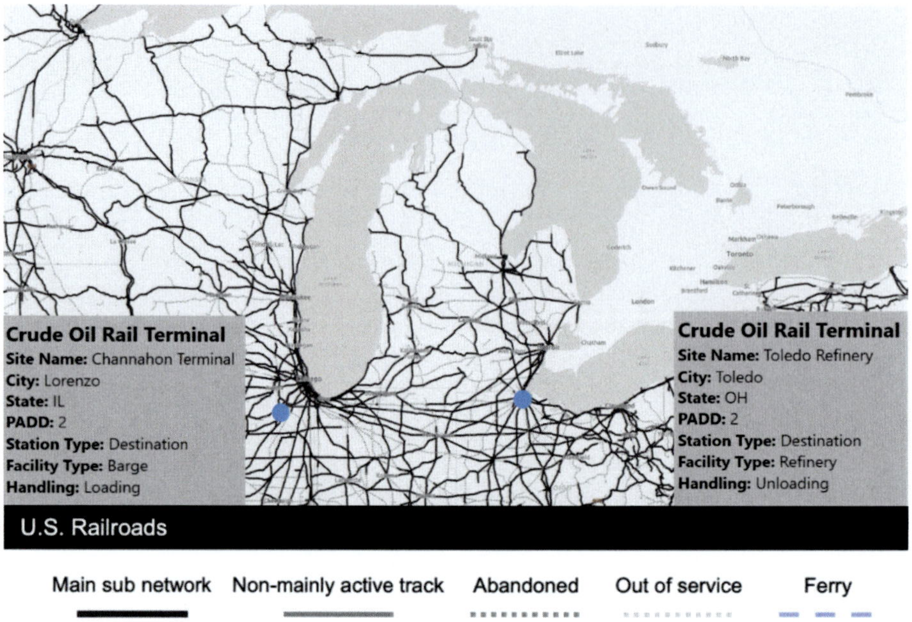

SOURCE: Created with U.S. Energy Information Administration, 2023.

Roads

The lowest overall volume of oil is moved via tanker truck.[47] Trucks are not a particularly efficient means of transport because a tanker truck can carry between 200 and 250 barrels of oil.[48] Trucks are generally used to move refined petroleum products from petroleum product terminals to other destinations.[49] Figure 2.5 shows the U.S. petroleum product terminals in the region.

---

[47] HAZMATON, "Crude Oil Transport by Truck in the Great Lakes," webpage, undated b.

[48] Congressional Research Service, 2014.

[49] HAZMATON, undated b.

Figure 2.5. U.S. Petroleum Product Terminals

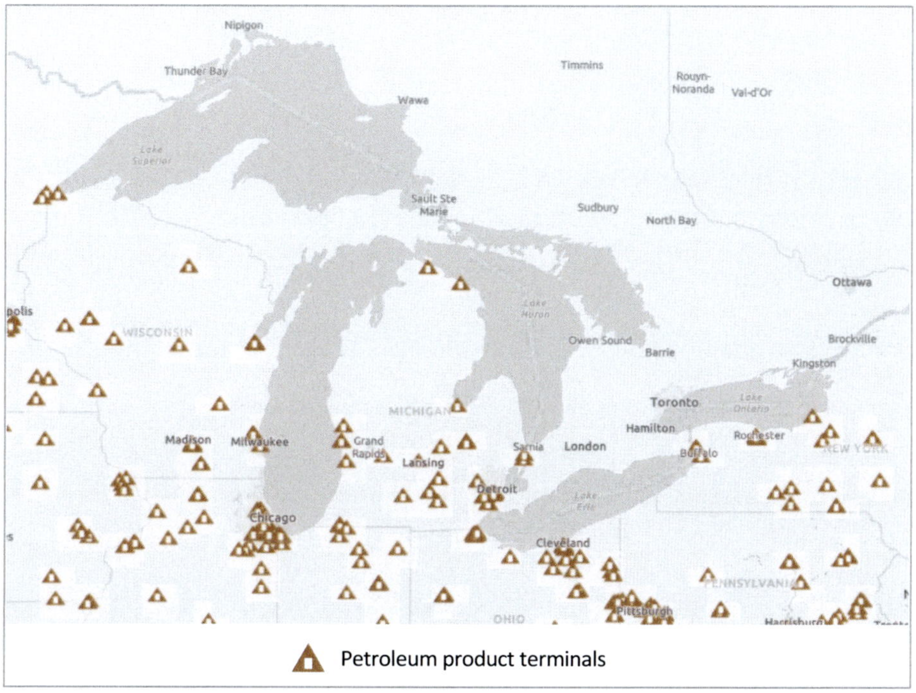

SOURCE: Created with U.S. Energy Information Administration, 2023.

Sources of Oil Spill Risk

Based on the identified locations and movements of oil discussed in the previous section, we then sought to determine areas of oil spill risk to the Great Lakes. In identifying potential risk areas, we considered the following oil sources:

- crude oil and petroleum product pipelines
- crude oil and petroleum product rail terminals
- petroleum refineries
- petroleum shipping waterways
- petroleum-based power plants
- U.S. ports
- U.S. and Canadian railroads
- oil wells
- oil tankers and other vessels by AIS tracks.

We collected geographic data on these oil risk sources from the Energy Information Administration; MarineCadastre.gov, a collaboration between the BOEM and NOAA; and NOAA's Environmental Response Management Application (ERMA). We restricted the data to oil spill risk sources within five miles of the Great Lakes coasts because they are the source that could potentially reach water and be within the USCG's jurisdiction. We aggregated the data into a set of geographic heatmaps, with a darker color representing a greater number of possible sources in that area. Figure 2.6 shows a map of all the U.S. oil spill risk sources that were considered.

Note that this data is purely geographic; we were unable to analyze the sources by likelihood of a spill or likely size of an oil spill event. The darkest color on these heatmaps shows the greatest number of potential oil spill sources, but any colored area has at least one possible source of an oil spill. The colored regions are centered on the actual source; the size of the circle around the source is not tied to the potential distance that oil could travel, but is purely for visualization purposes.

Figure 2.6. Heatmap of All Considered Oil Spill Risk Sources in the United States

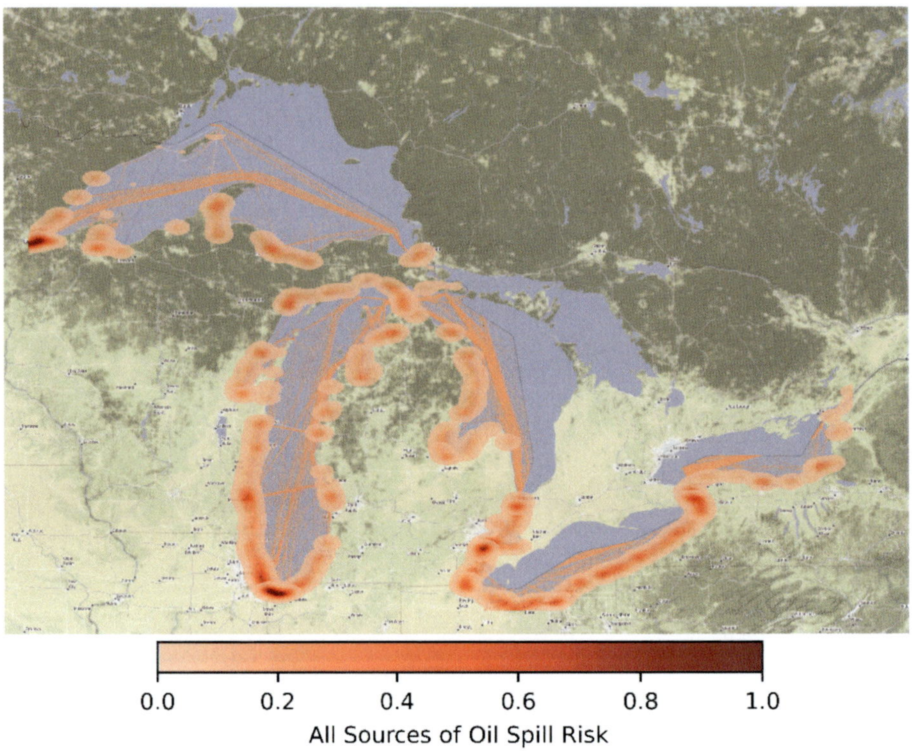

SOURCE: Features information from U.S. Energy Information Administration, "U.S. Energy Atlas," undated; NOAA and BOEM, undated.

We also categorized the data into U.S. nonstationary risk sources (Figure 2.7) and stationary risk sources (Figure 2.8) because response planning for each risk source might differ. Preparation for a rail car spill that could occur along any point of the railroad network, or a spill from a small vessel on Lake Michigan, for example, will differ from the preparation for a spill from a fixed location, such as an oil well. Stationary sources include the following:

- oil wells
- petroleum and crude oil pipelines
- petroleum power plants

- petroleum refineries
- petroleum and crude oil rail terminals.

Nonstationary sources include the following:

- railroads
- waterways through which oil is shipped
- ports[50]
- oil tanker and other vessel shipping lines by AIS tracks.

Figure 2.7. All Nonstationary Sources of Risk in the United States

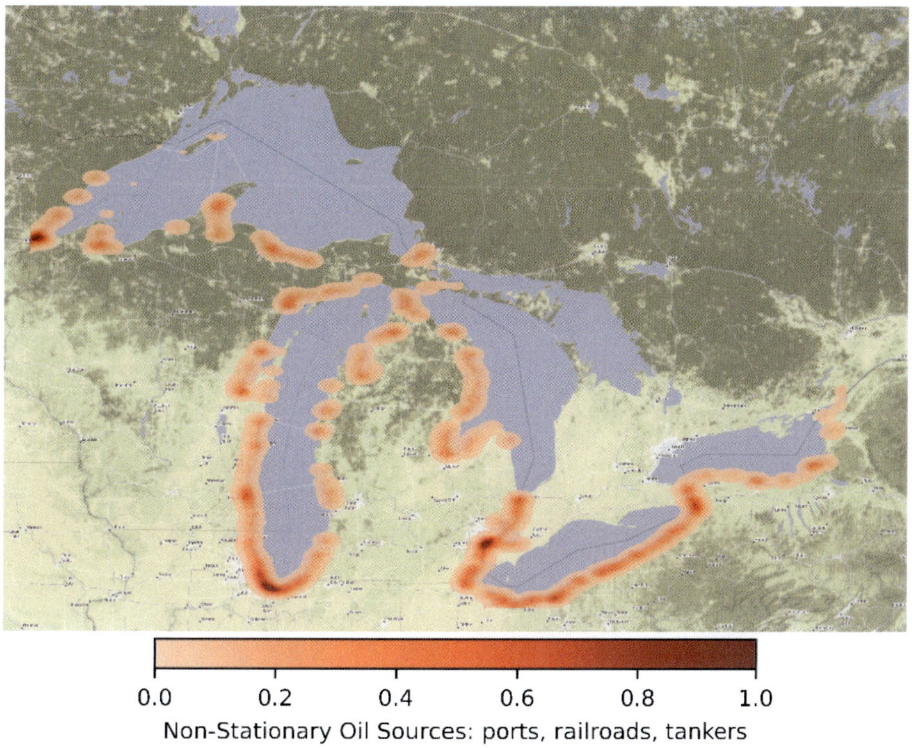

SOURCE: Features information from U.S. Energy Information Administration, undated; NOAA and BOEM, undated.

---

[50] Note that although ports are fixed locations, they are centers of activity for small vessels, a common source of small oil spills in the Great Lakes region. Unlike oil tanker tracks, there is limited data on small vessel traffic movements, so ports are used as a proxy to represent small vessel movements.

Figure 2.8. All Stationary Sources of Risk in the United States

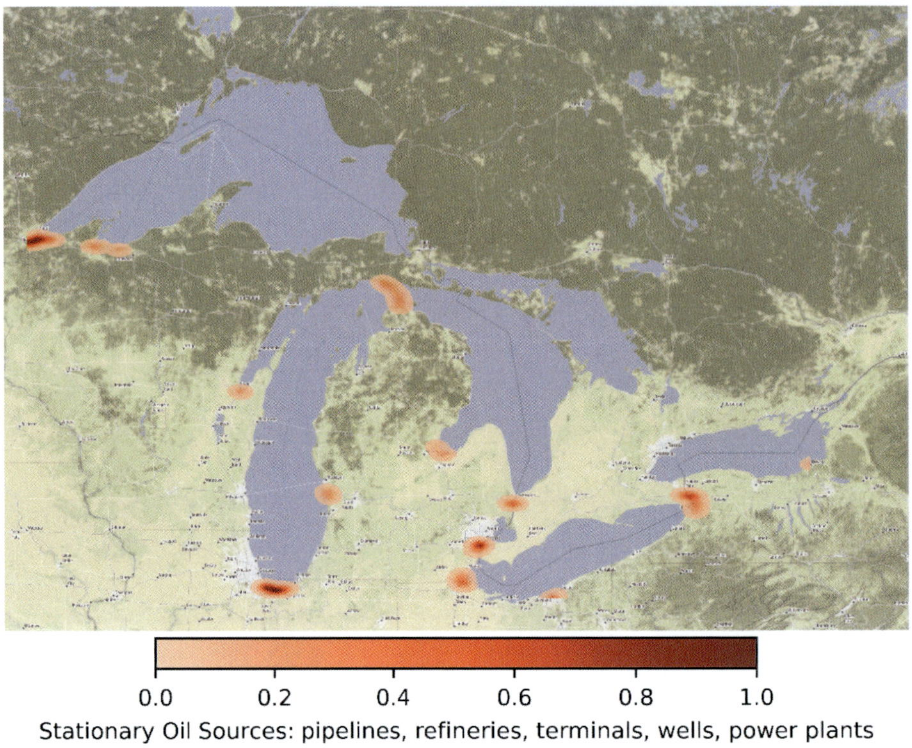

SOURCE: Features information from U.S. Energy Information Administration, undated.

Future Changes to Oil Flow in the Great Lakes Region

We identified a few notable ongoing issues that might affect how oil is moved through the Great Lakes Region in the future. Enbridge plans to replace part of Line 5, which moves crude oil through the region as part of Enbridge's Mainline System.[51] Part of Line 5 sits on the lakebed under the Straits of Mackinac, and the replacement project would tunnel through the lakebed to enclose the pipeline.[52] In 2020, Michigan Governor Gretchen Whitmer ordered Enbridge to cease

[51] Enbridge Inc. offers crude oil and liquids pipeline systems for energy projects, natural gas transmission, and midstream businesses, and distributes electricity and retail energy products (Bloomberg, "Enbridge," company profile, undated). Enbridge moves approximately 30 percent of the crude oil produced in North America, transports nearly 20 percent of the natural gas consumed in the United States, and operates North America's third-largest natural gas utility by consumer count. Headquartered in Calgary, Canada, Enbridge operates the world's longest and most complex crude oil and liquids transportation system, with approximately 17,809 miles of active pipeline across North America (9,299 miles in the United States and 8,510 miles in Canada (Enbridge, "Liquids Pipelines," webpage, undated b).

[52] Annette Giachino, "Enbridge Provides Update on Tunnel Project," *Upper Michigan's Source*, February 20, 2023; Enbridge, *Investing in Michigan's Future*, March 2022.

operations in the Straits of Mackinac,[53] but Enbridge has continued to operate in the area and is seeking permits for the project.[54] In addition, a 12-mile section of Line 5 runs through the territory of the Bad River Band of the Lake Superior Chippewa.[55] Enbridge has been ordered to cease operations in that territory no later than June 16, 2026, and is working to acquire permits to reroute that section of pipeline.[56] If the pipeline is not rerouted by that date, Line 5 will effectively be shut down. [57] A shutdown of Line 5 would effect crude oil movements in the region and, assuming oil demand does not decrease, would increase demand for oil movements from other sources (e.g., rail).

Conclusion

In conclusion, the Great Lakes region contains numerous oil wells and petroleum refineries and is a transportation route for a large volume of oil, with pipelines being the primary means of transport. Crude oil is not shipped on the Great Lakes, but a variety of petroleum products are shipped to or from Great Lakes ports. However, the fate of Enbridge's replacement of Line 5 could change the calculus for how oil moves through the region.

---

[53] Gretchen Whitmer, "Governor Whitmer Takes Action to Shut Down the Line 5 Dual Pipelines Through the Straits of Mackinac After a Reasonable Transition Period to Protect the State's Energy Needs," press release, State of Michigan Office of the Governor, November 13, 2020.

[54] Leana Hosea, "Why Michigan is Trying to Shut Down Canada's Enbridge Line 5 Pipeline," BBC, January 15, 2023; Giachino, 2023; Enbridge, 2022.

[55] Enbridge, "Line 5 Through the Bad River Reservation," undated.

[56] James McCarten, "Enbridge 'Must Cease' Line 5 Operations on Bad River Band Territory by June 2026: Judge," The Canadian Press, June 27, 2023.

[57] Sheri McWhirter, "Wisconsin Line 5 Shutdown Order May Disrupt Pipeline Fight in Michigan," Michigan Live, June 21, 2023.

Chapter 3. Response Capabilities and Capacities of Government Agencies and Private Companies on Both Sides of the Border

This chapter is divided into two main sections. The first discusses the roles and responsibilities of the different agencies, responders, and owners and operators for spills involving the United States and then for spills that cross the border. The second examines a variety of oil-spill response techniques, explaining how they work, their shortcomings, and how the United States and Canada currently employ them. Because response techniques differ depending on the environment where they are used, we split the discussion into techniques for fresh water and then those for icy conditions. More detail on this discussion can be found in Appendix A.

The United States and Canada have a formal structure for cooperating on oil-spill response in the Great Lakes. The Canada-U.S. Joint Marine Pollution Contingency Plan lays out the procedures for preparedness and response to oil and other pollutants that cross the maritime boundary in the Great Lakes in fulfillment of the requirements laid out in the International Convention on Oil Pollution Prevention Response and Cooperation of 1990. Each region has an annex in the plan that provides detailed procedures for response planning, operations, and waste management. The Great Lakes annex, called the CANUSLAK, covers six states, one province, three EPA regions (2, 3, and 5), and several indigenous communities and tribes. The CANUSLAK provides the necessary information to respond in a specific geographic area and must be reviewed annually for legal, policy, environmental, organizational, or socioeconomic changes in the region. The Joint Response Team integrates the response capabilities of the United States and Canada and is cochaired by the USCG District Incident Management and Preparedness Advisor and the Canadian Coast Guard (CCG) Regional Director of Programs.[58] The phases in the joint contingency response are shown in Figure 3.1.

Figure 3.1. Response Phases Identified in the Joint Contingency Plan

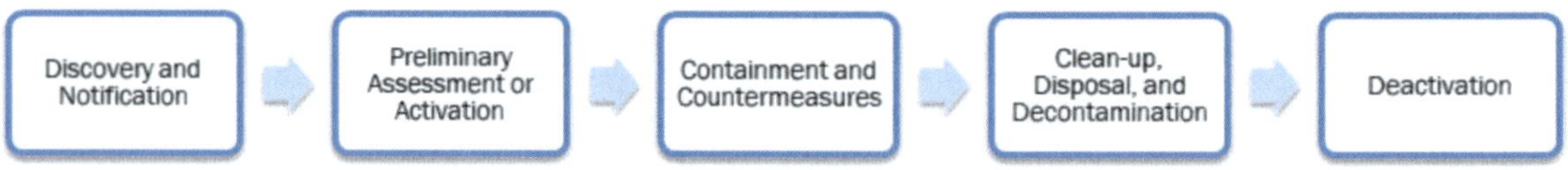

[58] *Canada–United States Joint Contingency Plan*, Canadian Coast Guard and United States Coast Guard, August 3, 2017; *Annex 1, Great Lakes Operational Supplement to the Joint Marine Pollution Contingency Plan*, Canadian Coast Guard and United States Coast Guard, 2022.

Roles and Responsibilities

First, we discuss oil spills that involve the United States, then we discuss spills that cross the U.S.-Canadian border, and finally, we note the differences between the two.

Spills that Involve the United States

The roles and responsibilities of U.S. government organizations for oil-spill response are clearly laid out in 33 C.F.R. 154 subparts F, H, and I. Oil-spill response has a multilayered, multiorganizational structure. Oil-spill response funding is provided by private owners and operators. The U.S. government's role is oversight of cleanup activities, but it is not intended to be the primary responder to spills. However, there are times—when spills are large in scope, large in risk to the public or environment, or a responsible party has not been identified—when the federal government must take on a large role or total responsibility for the spill response efforts. This is also true if the responsible party is determined to be taking insufficient action.

The federal organizations involved in spill response depend on the size and scope of the spill. For all spills, regardless of size, the National Response Center (NRC), which is staffed by the USCG, is informed of the spill via its hotline.[59] Once the NRC is notified, a Federal On Scene Coordinator (FOSC) is assigned; the EPA is the FOSC for inland zone spills and the USCG is the FOSC for on-water and coastal zone spills.[60] The designation of inland versus coastal regions is not a line on a map; instead, it varies from region to region and can also vary based on the spill circumstances.

For most spills, the Captain of the Port (or the Sector Commander) has authority over response oversight, as FOSC. In the Great Lakes, there are four sectors: Sector Sault Sainte Marie, Sector Lake Michigan, Sector Detroit, and Sector Buffalo, and one Marine Safety Unit in Duluth, Michigan. If the Sector Commander requests assistance, the Great Lakes are all housed under USCG District 9, which can provide additional support in the form of a District Response Advisory Team (DRAT) and additional equipment.[61]

If additional federal support is needed, the USCG maintains a National Strike Force (NSF), which is a deployable specialized force that provides "rapid response support for incident management, site safety, contractor performance monitoring, resource documentation, response strategies, hazard assessment, oil spill dispersant and in-situ burn (ISB) use, operational effectiveness monitoring, and high capacity lightering and offshore skimming capabilities."[62]

---

[59] USCG, *Great Lakes Oil Spill Response and Cleanup Activities Assessment, Report to Congress*, November 21, 2018a.

[60] USCG, 2018a.

[61] USCG, 2018a.

[62] USCG, 2018a.

It should be noted that these federal resources are all available, but there is no designated time when escalating the incident must occur. Choosing to request DRAT or NSF support is the decision of the USCG FOSC, and past support has been in the form of air support and personnel support for longer-term spill response oversight.[63]

A variety of other federal, state, local, and tribal organizations can be involved in oil-spill response efforts. The EPA and local municipalities monitor and protect drinking water. EPA regions 2, 3, and 5 cover the Great Lakes region.[64] NOAA serves as the Scientific Support Coordinator, which provides modeling, baseline and applied research, mapping, forecasting, and survey support.[65] Additional federal support might be provided by the Department of Transportation's Pipeline and Hazardous Materials Safety Administration, Centers for Disease Control and Prevention, Department of Labor, Occupational Safety and Health Administration, Army Corps of Engineers, and DHS, as needed.[66]

Owner and Operator Responsibilities

Despite the well-mapped out federal roles and responsibilities for oil-spill response, vessel and facility owners and operators are the primary parties responsible for response and cleanup efforts. Facility and vessel owners and operators are required to develop response plans, which are reviewed and approved by the USCG and/or the EPA. Most owners and operators develop response plans that utilize Oil Spill Response Organizations (OSROs), which are private companies that are contracted to clean up after a spill. Plans that receive USCG approval are valid for up to five years,[67] during which time the plan holder is required to review its vessel response plan (VRP) annually and resubmit it for review and approval if there are any changes.[68] OSRO information is reported to the USCG's NSF Coordination Center and is maintained in a Response Resource Inventory database. Oversight of OSROs is a USCG responsibility, and OSRO qualifications are well defined in the *Guidelines for the U.S. Coast Guard Oil Spill Removal Organization Classification Program*.[69]

It should be noted that although the USCG does own and maintain some locally held spill response equipment, it is not legally responsible for providing spill response

---

63 USCG interview with authors, February 9, 2023.

64 EPA, "Regional and Geographic Offices," webpage, updated January 17, 2023.

65 USCG, 2018a.

66 USCG, DHS, 2018.

67 Title 33, Chapter 1, Subchapter O, Part 155, Subpart D, Tank Vessel Response Plans Oil, Section 155.1065 (c).

68 According to USCG regulations, some revisions or amendments to an approved VRP must be submitted to the VRP Program for approval, such as a change in the vessel owner or operator, a significant change in vessel configuration, or changes to emergency response procedures (GAO, *Coast Guard: Improved Analysis of Vessel Response Plan Use Could Help Mitigate Marine Pollution Risk*, GAO 20-554, September 29, 2020, p. 9).

69 USCG, *Guidelines for the U.S. Coast Guard Oil Spill Removal Organization Classification Program*, December 2021d.

equipment or personnel. OSROs are the main holders of spill response equipment in the Great Lakes, and owners and operators hold the legal responsibility for oil-spill response.

Spills That Involve Canada

As in the United States, vessel owners and oil handling facilities are responsible for oil spills in Canada, but Canada has a slightly different spill response structure. Canada's federal transportation department, Transport Canada, has overall responsibility for regulatory requirements and response guidelines and standards. Responsible parties are required to have the capacity to respond to spills of up to 10,000 tons and to deploy equipment on-site within 72 hours.[70] Vessels and oil handling facilities are required to have an agreement with a response organization, and Transport Canada certifies OSRO capabilities on a three-year basis. The ECRC is the only certified response organization for the Great Lakes region in Canada.[71] It provides response services to ships and oil-handling facilities in accordance with Canadian law.

In Canada the spill source, not the medium for the spill, determines the lead agency that oversees the responsible party's response operations. The CCG coordinates response operations for oil spills that originate from vessels or mystery sources that affect waterbodies, and it acts as the on-scene commander if a responsible party is not identified. Other Canadian agencies are responsible for spills that originate from pipelines (Canada Energy Regulator) or on land (the provincial government, which in the Great Lakes region is the Ontario Ministry of the Environment, Conservation and Parks).[72] Two agencies can provide scientific information to spill response planning and operations. Environment and Climate Canada is the lead agency for providing scientific information on environmental and wildlife issues and is responsible for maintaining the National Environmental Emergencies Management System that has geospatial data on sensitive areas. However, the utility of this data is questioned, in part because some data sets are outdated and in part because data sharing agreements do not cover emergency response.[73] Environment Canada with Fisheries and Oceans Canada provides information on spill trajectory and dispersion, recovery and cleanup techniques, forecasts of weather forecasts

---

[70] Tier 1 spills of 150 tons require response times of 6 hours, tier 2 are 1,000 tons within 12 hours, and tier 3 are 2,500 tons within 18 hours.

[71] "ECRC is a private management company, owned by several of the major Canadian oil companies, whose role is to provide marine oil spill response services, when requested, to the 'responsible party, the Canadian Coast Guard or to any other Government Lead Agency." "ECRC (operating as ECRC-SIMEC) is certified by Transport Canada–Marine Safety as a response organization under the Canada Shipping Act (CSA)." Eastern Canada Response Corporation, "About ECRC," webpage, undated.

[72] Transport Canada, *A Review of Canada's Ship-Source Oil Spill Preparedness and Response Regime: Setting the Course for the Future*, Tanker Safety Panel Secretariat, 2013, p. 19; Annex 1, 2022, p. 9; Naznin Sultana Daisy, Mohammad Hesam Hafezi, Lei Liu, and Kenneth Lee, "A Comprehensive Review of Canadian Marine Oil Spill Response System Through the Lens of Decanting Regulations and Practices," *Journal of Marine Science and Engineering*, Vol. 10, No. 9, 2022; interview with scientist, March 8, 2023.

[73] Transport Canada, 2013, p. 36; Canadian interviewee February 27, 2023.

and lake conditions (including ice), and location of sensitive areas; it can also provide scientific equipment and assessments.[74]

Differences Between U.S. and Canadian Roles and Responsibilities

In both the United States and Canada, oil-spill response and cleanup is the responsibility of vessel operators and facility owners, and comparable federal agencies have similar responsibilities with a few exceptions. For one, in U.S. waters, the USCG is responsible for providing regulatory guidance, oversight, and operational coordination regardless of the spill source. In contrast, the CCG is responsible for operational oversight and coordination only if a spill originates from a vessel, or an unknown source, but relies on Transport Canada for vessel information and oversight of response capabilities. Another difference is that vessel operators in the United States are required to develop a vessel response plan that is certified by the USCG and provides information on the capabilities available to manage oil spills to meet the requirements of 33 C.F.R. Part 155. Canadian requirements are the same for all regions and are limited to spills of 10,000 tons or less.[75] Finally, U.S. vessel operators might maintain in-house response and recovery capabilities or contract with OSROs, while Canadian response capabilities are provided by one OSRO in the Great Lakes Region—the ECRC, a public-private entity that is certified by Transport Canada.

Capabilities and Capacities in Oil-Spill Response

This section describes three types of oil-spill response techniques—monitoring, mechanical, and nonmechanical—and how the United States and Canada currently employ them. We start with response techniques for fresh water and then shift to techniques for icy conditions.

When oil is spilled into a body of water, natural processes such as evaporation and biodegradation reduce the amount of oil in the environment.[76] These natural processes might be sufficient to mitigate very small oil spills, but the quantity and type of oil spilled could require human intervention for proper cleanup. Response entities can deploy a variety of techniques to contain and clean up an oil spill, and the optimal response will depend on a host of factors, including but not limited to the type and quantity of oil spilled, the source and location of the spill, and the weather. Based on this information, response entities will select one or, more likely, a combination of techniques appropriate for the spill as informed by current guidance.[77] Quickly

---

[74] Transport Canada, 2013, p. 36.

[75] Transport Canada, 2013, p. 14.

[76] International Tanker Owners Pollution Federation Limited, *Fate of Marine Oil Spills*, Technical Information Paper 2, 2014a.

[77] In the United States, the NCP is the primary federal policy for directing oil and hazardous substance spill response. The NCP informs Regional Contingency Plans (RCPs), which in turn inform Area Contingency Plans

deploying an appropriate response to an oil spill is imperative for mitigating environmental harm.

The fate of oil can be different in marine and fresh water environments because of water density and salinity considerations. Spills of heavier oils in fresh water environments might necessitate a different response capability or strategy than a salt water environment. Heavier oils that would normally float in the greater density salt water might sink to the bottom or drift in the fresh water column, complicating the search for and access to the spilled oil.[78]

Emerging Research

In a 2018 report to Congress, the USCG acknowledged that "the understanding of the behavior of oil in fresh water environments is largely driven by science and empirical evidence that was gathered during large salt water spills."[79] As a result, there is a lack of research examining the nuances of oil-spill response in fresh water environments. **Although conventional oil-spill response techniques can be deployed in both marine and fresh water environments, these techniques are primarily used in salt water and might not be optimized for response in the unique environment of the Great Lakes.**

Capabilities and Capacities in Normal Freshwater Conditions

Monitoring Equipment

Monitoring equipment provides response entities with valuable information, such as the size of the oil spill, the thickness of the slick, and the movement of the oil, which can inform proper response planning and execution. Monitoring equipment is particularly important when oil is spilled in hard-to-observe areas, such as beneath ice, as we discuss in a later section. Monitoring equipment that can be deployed on the Great Lakes includes unmanned surface vehicles (USVs), unmanned underwater vehicles (UUVs), unmanned aerial vehicles (UAVs), and remote sensors.

Unmanned vehicles allow responders to surveil an oil spill in areas that are relatively inaccessible to humans. Autonomous or remote-controlled UAVs can capture aerial views that can help determine the size and movement of the spill, identify obstructions that could impede response, and determine the proximity of the spill to sensitive areas.[80] Similarly, USVs and

---

(ACPs). The RCPs and ACPs are specific to focused areas of operation and direct agencies on how to respond to an oil spill in that area. USCG, DHS, 2018.

[78] NOAA, Office of Response and Restoration, "How Is an Oil Spill in a River Different Than One in the Ocean?" webpage, July 31, 2015.

[79] USCG, DHS, 2018.

[80] Interagency Coordinating Committee on Oil Pollution Research (ICCOPR), *Oil Pollution Research and Technology Plan: Fiscal Years 2022–2027*, December 2021.

UUVs can identify, track, and sample oil in subsurface operations.[81] UAV technology is more mature and widely utilized in overhead observation, while USVs and UUVs are less developed and employed but can be capable of operating in an oil slick and possibly in ice.

Unmanned vehicles can be equipped with remote sensing devices to locate and observe oil during a spill. These sensing technologies can also be used in manned vehicles or operated manually. Importantly, there are many emerging remote sensing technologies that could potentially be used to observe an oil spill under circumstances that might normally prohibit human observation. Various types of radar and infrared cameras and other systems such as laser fluorosensors and light detecting and ranging can be used to monitor an oil spill and inform the appropriate response.[82]

Mechanical Response in Normal Freshwater Conditions

Once an oil spill has been identified, response operations commence. Mechanical response techniques are the primary method for responding to oil spills in the United States.[83] Mechanical responses remove the oil from the water without introducing foreign substances into the environment.[84]

Typically, the first step in mechanical response operations is containing the spread of the oil using booms. Booms are long pieces of equipment that float on water and are positioned around an oil slick to prevent the oil from spreading. Because oil is generally less dense than water, oil will float on the water's surface and the booms will contain the slick.[85] This containment helps mechanical response techniques operate more effectively. Booms can be used to both contain and concentrate the oil and divert it toward a location more suitable for recovery operations or away from a sensitive area.[86] Figure 3.2 shows a worker helping to position a boom in the water.

---

[81] USVs and UUVs describe systems that are unmanned and specify the operation location of the vehicle (i.e., on or below the water's surface). These vehicles are sometimes referred to as autonomous underwater vehicles (AUVs) or remotely operated vehicles. The designators "autonomous" and "remote operated" describe the necessary degree of human control to operate the vehicle. Both USVs and UUVs can be autonomous, remote operated, or some hybrid of the two. ICCOPR, 2021.

[82] Robert K. Watkins, and Alan A. Allen, "Remote Sensing Guide to Oil Spill Detection in Ice-Covered Waters," Arctic Response Technology, June 12, 2016.

[83] EPA, "EPA's Response Techniques," webpage, updated December 2022f.

[84] This contrasts with a response option such as the use of *chemical dispersants*, which are chemical agents that are mixed with, and remain in, the water. This response option is described in more detail in Appendix A.

[85] The type of oil spilled, the salinity of the water, and the weather conditions will significantly affect how much of the spilled oil floats on the surface and how much sinks into the water column. *Diluted bitumen* is crude oil with a specific gravity, which causes it to sink in fresh water, making recovery more difficult (S. Stoyanovich, J. R. Rodríguez-Gil, M. L. Hanson, B. P. Hollebone, D. M. Orihel, V. P. Palace, R. Faragher, F. S. Mirnaghi, K. Shah, Z. Yang, and J. M. Blais, "Simulating Diluted Bitumen Spills in Boreal Lake Limnocorrals: Part 2–Factors Affecting the Physical Characteristics and Submergence of Diluted Bitumen," *Science of The Total Environment*, Vol. 790, 2021).

[86] International Tanker Owners Pollution Federation Limited, *Use of Booms in Oil Pollution Response*, Technical Information Paper 3, 2011.

Figure 3.2. Example Oil Boom

SOURCE: U.S. Navy photo by Patrick Nichols, ID 100504-N-6268N-073, 2010.

Once the oil is contained, skimmers can be used to recover the spilled oil. *Skimmer* is a broad term for a device that collects oil from the water and transports the oil and water mixture into a tank for oil recycling or disposal. Skimmers can be self-propelled or operated from a vessel or from shore.[87] The EPA categorizes skimmers into three types: Weir skimmers, oleophilic (oil-attracting) skimmers, and suction skimmers.[88]

Weir skimmers use a dam or enclosure device that allows oil to flow over the dam and into a well. This method minimizes the amount of water that is recovered with the oil. Once inside the well, the oil can be pumped to the storage tank, where it is recycled or disposed of.[89]

Oleophilic skimmers "use belts, disks, or continuous mop chains of oleophilic materials to blot the oil from the water surface. The oil is then squeezed out or scraped off into a recovery tank," according to the EPA.[90] Recovered oil can then be recycled or disposed of.

Suction skimmers operate like a vacuum. The skimmer head sucks up oil floating on the water's surface and pumps it into a storage tank for recycling or disposal.[91]

Skimmers are often a preferred response option because they have little to no effect on the environment and the recovered oil can be recycled. However, the overall effectiveness of skimmers is up for debate. One NOAA estimate posits that under ideal conditions (i.e., a

---

[87] EPA, "Skimmers," updated September 2022c.

[88] EPA, 2022c.

[89] EPA, 2022c.

[90] EPA, 2022c.

[91] EPA, 2022c.

sufficiently thick slick and calm water conditions relatively free from debris), mechanical recovery can recover about 40 percent of spilled oil. NOAA also estimates that during the Deepwater Horizon response, skimmers only managed to recover about 3 percent of spilled oil.[92]

Sorbents are oil-attracting and water-repellant materials that can absorb or hold oil, allowing it to be removed from the water with the sorbent material. Sorbents typically come in three categories: natural organic (e.g., peat moss, wool, straw, saw dust), natural inorganic (e.g., sand, clay, ash), and synthetic (e.g., polyurethane, polypropylene).[93] The sorbents can be used on water, often in the form of a sorbent boom, or used on or near shore in the form of an absorbent pad.

On or near shore operations might require or allow for response techniques that are not viable on open water. When spills occur on the shore, responders can manually or mechanically remove the oil using shovels or backhoes. It might also be necessary to use surfactants and hoses to clean oil from the shoreline or rip-rap, which allows the oil to collect in the water, where it can be removed using sorbents, skimmers, and other means.[94]

Nonmechanical Response in Normal Freshwater Conditions

Mechanical recovery is the primary response to oil spills, but nonmechanical techniques can also be used. Although these options might be effective, they affect the environment in ways that mechanical techniques do not. Understanding the costs and benefits of these techniques is particularly important in the Great Lakes because many people live in the region and get their drinking water from the lakes. This is in contrast to responding to oil far out at sea where adverse effects on humans are less of a concern.

In-Situ Burning

ISB involves igniting the oil slick, allowing the oil to burn, and monitoring the burn as it occurs. The thickness of the oil slick largely determines the efficiency of the burn (along with environmental factors such as ambient temperature and wave action).[95] An ISB removes the oil from the water but produces emissions and burn residue. The emissions can contain dangerous particles that can be harmful to humans, so it is important to understand the likely quantity of emissions that will be produced from a burn and to continually measure the air quality downwind from the burn. Although burn residue does not pose an acute aquatic toxicity concern, the residue can sink and smother organisms on the floor of the body of water.[96]

---

[92] NOAA, Office of Response and Restoration, "What Have We Learned About Using Dispersants During the Next Big Oil Spill?" April 2015, updated November 2022.

[93] ICCOPR, 2021.

[94] NOAA, National Ocean Service, "How Does NOAA Help Clean Up Oil and Chemical Spills?" undated b.

[95] ICCOPR, 2021.

[96] NOAA, Office of Response and Restoration, "Residues from In Situ Burning of Oil on Water," February 2019.

Chemical Agents

The EPA maintains a list of chemical agents approved for use in oil-spill response. These agents, found within the NCP Product Schedule, must meet effectiveness and toxicity standards. RCPs and ACPs can preauthorize chemical agents for use provided they are approved on the NCP product schedule. However, most chemical agents are not pre-authorized for use in the Great Lakes by either the United States (via RCPs or ACPs)[97] or Canadian authorities.[98] Ultimately, the RCPs and ACPs will contain a blueprint for spill response specific to the area, and the lead agency will have some freedom to choose from approved response techniques to employ. In the next sections, we discuss various types of chemical agents that are part of nonmechanical responses to oil spills.

Chemical Dispersants

Chemical dispersants are chemical agents that can break oil particles down into smaller droplets that disperse in the water column. The dispersants break up the surface oil slick, which allows smaller participles to spread throughout the body of water, making it easier for natural processes to break down the oil particles.[99] Chemical dispersants can be applied over a large area by aircraft, over a more focused area by a surface vessel, or beneath the surface by a UUV.[100] The type of oil, ambient temperature, water conditions, and chosen dispersant will all determine the effectiveness of the dispersant's use. It is also recommended that dispersants be used early in the response process because oil becomes more viscous as it weathers, which makes it more difficult for the dispersants to break down.[101]

Solidifiers

Solidifiers are chemical agents that react with oil to form a solid substance. Once the reaction occurs, the gelatin-like solid can be removed from the water using mechanical means. A significant ratio of solidifier to oil—sometimes up to 3:1—must be used to be effective. Depending on the size of the spill and water conditions, it can be prohibitively difficult to remove the solid product from the water and store and dispose of it.[102]

---

[97] USCG, 2015, p. 32. All other ACPs we reviewed have similar language pertaining to this topic.

[98] LimnoTech, *Impacts of Unrefined Liquid Hydrocarbons on Water Quality and Aquatic Ecosystems of the Great Lakes Basin*, January 2018, p. 57.

[99] EPA, "Dispersants," updated November 2022f.

[100] GAO, *Offshore Oil Spills: Additional Information is Needed to Better Understand the Environmental Tradeoffs of Using Chemical Dispersants*, GAO-22-104153, December 2021.

[101] GAO, 2021.

[102] ICCOPR, 2021.

Herders

Herders are surfactants (i.e., a substance that reduces surface tension) that mix with oil to herd oil slicks toward or away from a particular area, or to direct two or more separate oil slicks to merge.[103] Merging slicks can be useful because thicker oil slicks are easier to recover through mechanical means (as the thick slick is easier to separate from the water) and through ISB (because thicker slicks are easier to ignite and keep burning).

Capabilities and Capacities in Icy Environments

The Great Lakes region has varying levels of ice coverage,[104] which can make oil-spill response much more complicated. Ice can make it difficult for responders in vessels to reach an affected area, and oil spilled beneath ice can be very difficult to identify and reach. Additionally, freezing temperatures and less sunlight during the winter can affect equipment and make operations more difficult and dangerous. However, ice coverage can also help contain the spread of spilled oil. The ice coverage level, type of ice, and spill type (i.e., on, under, or in ice) will determine the type and efficiency of response operations.

Monitoring Equipment in Icy Environments

In icy conditions, unmanned vehicles and remote sensors might be the only way for responders to see the oil spill. If there is enough ice to impede vessel movement or the oil is below sheets of ice, then monitoring equipment can be used to identify the spill location and inform responders where to cut into the ice to allow the oil to be recovered.

Mechanical Response in Icy Environments

When oil is underneath or contained in ice, responders must first expose the oil before beginning response operations. This can be done by cutting sections of the ice with a chainsaw or other tool to create a space where the oil can collect in an exposed area. Once the oil is exposed, mechanical or other response techniques can be employed. The ice thickness will dictate how responders expose the oil. Sufficiently thick sheets of ice will support vehicles or people who can cut into the sheets.[105] Total ice coverage by thin sheets could call for vessel-mounted equipment to break the ice and expose the oil.

Even if the spill area is not 100 percent covered with ice, partial coverage can still significantly impede response. Booms can still be deployed to contain oil from spreading, but large chunks of ice can damage or degrade the booms. Additionally, the boomed area will

---

[103] ICCOPR, 2021.

[104] Coverage can vary from 0 percent to 100 percent ice cover in an area depending on the lake, season and year (NOAA, Great Lakes Environmental Research Laboratory, "Great Lakes Ice Cover," webpage, March 2023.

[105] One interviewee noted their organization's use of ice-cutting sleds. A blade or saw mounted on the sled cuts into the ice as the sled is maneuvered, as shown in Figure 6.1. Eastern Canada Response Corporation member, phone interview with the authors, April 21, 2023.

contain not only oil but also ice chunks, which will need to be cleaned of oil. Skimmers can be deployed in icy environments, but the weir and suction skimmers are unlikely to be very effective because they are susceptible to clogs by debris. Oleophilic skimmers are likely the best skimmer option in partially icy environments.[106]

Nonmechanical Response in Icy Environments

Like mechanical response techniques, nonmechanical methods are also less effective in icy conditions. The degree of ice coverage would have the largest influence on the effectiveness of chemical agents in icy environments. Dispersants need to reach and mix with the oil, so if the ice coverage prevents or inhibits this process, dispersants will not be as successful in breaking down oil particles. Similarly, sorbents and herders need to mix with the oil, which is complicated in icy environments, and their effectiveness could be affected by severe temperatures.

Although an ISB response still requires access to the oil, field and lab tests have shown that ISBs can work well in icy environments. When there is significant ice coverage (around 70 percent) the ice can act as natural containment, which could allow burns to occur without the use of booms. In low ice coverage areas (around 30 percent) normal ISB techniques can also be effective. Ice coverages in between these levels could inhibit burn effectiveness without providing the natural containment advantage.[107] Finally, cold temperatures could affect the ignition temperature of some oils making it difficult or impossible to ignite the oil slick.

Conclusion

Response entities can deploy a variety of techniques to contain and clean up an oil spill, and the optimal response will depend on a host of factors, including but not limited to the type and quantity of oil spilled, the source and location of the spill, and the weather. As discussed above, although conventional oil-spill response techniques can be deployed in both marine and fresh water environments, these techniques are primarily used in salt water and might not be optimized for response in the unique environment of the Great Lakes.

Icy environments create additional challenges for oil spill responders. The cold weather can degrade human and equipment performance, frozen waterways can hinder or prevent response vessel maneuverability, and ice becomes an additional obstacle and safety hazard that must be accounted for in the response. Specialized monitoring equipment can be used to identify and track the spill in circumstances where human visual confirmation is not feasible. Responders must then determine the best way to access the oil and this will be dependent on the ice coverage and thickness. Once responders can access the oil, mechanical or nonmechanical response

---

[106] J. Wilkinson, C. J. Beegle-Krause, K-U. Evers, N. Hughes, A. Lewis, M. Reed, and P. Wadhams, "Oil Spill Response Capabilities and Technologies for Ice-Covered Arctic Marine Waters: A Review of Recent Developments and Established Practices," *Ambio*, Vol. 46, Supp. 3, December 2017.

[107] Wilkinson et al., 2017.

techniques can be employed. The type of technique used should consider how the specific kind of ice will affect the response.

Chapter 4. Oil Spill Response Organization Response Capabilities Analysis

The ability of oil-spill response organizations to get oil spill cleanup equipment to remote areas was a concern that arose in our interviews.[108] This chapter looks at these concerns analytically to see whether existing data support them. The method used finds the extent to which Great Lakes OSROs can respond to areas of concern in the Great Lakes. The motivation of this is to find where in the Great Lakes there could be a vulnerability in response, where OSROs are not able to get enough cleanup equipment on site in the required amount of time.

Regulations list required response times for OSROs to reach oil spills from different sources, in different areas. Title 33 § 154.1045(f), facility categorizes spills in the Great Lakes into Tiers 1 through 3.[109] The lowest tier, Tier 1, has the shortest required response time of 12 hours. We analyzed what OSROs could reach the priority sites within this shortest time frame. To help visualize where the OSROs are, Figures 4.1, 4.2, and 4.3 show the approximate locations of where OSROs have oil spill cleanup equipment in the Great Lakes region.[110] Figure 4.2 overlays historic oil spills on top of those locations. Figure 4.3 shows OSRO locations overlayed with a heatmap of where there is risk of possible oil spills from pipelines, refineries, railroads and rail terminals, oil wells, ports, shipping lines, and waterways through which oil is shipped. These three maps help to visualize where response planners might want to prioritize efforts or more closely validate the potential sources, response capabilities, and overall spill risk. For example, Figure 4.3 points to some potential response gaps along the south, central coast of Lake Superior, northeast coast of Lake Michigan, and along the southeasterly coast of Lake Erie. Figures presented in this analysis do not include historical spills exclusively within Canadian territory.

To see where in the Great Lakes OSROs might take too long to reach, we looked at the priority response sites listed in the response plans of the four sectors—sectors Buffalo, Detroit, Lake Michigan, and Sault Sainte Marie—and USCG's Marine Safety Unit Duluth. Each priority site is an area that must be blocked off with an exclusion boom in the event of a spill to prevent oil from reaching a critical area, like an endangered habitat or drinking water intake. The response plans that include these sites list the minimum boom length that is required to fully block off the area. It is important to note that this is only what is needed to block off a critical area with an exclusion boom, not the boom required to contain the spill itself.

---

[108] Interview with responsible party and response organization, March 8, 2023, and April 25, 2023, respectively.

[109] Title 33, Chapter I, Subchapter O, Part 154, Facilities Transferring Oil or Hazardous Material in Bulk.

[110] The locations are approximate because the source we received the exact data from is FOUO. To protect that information, the locations in Figures 4.1–4.3 show the townships that have one or more OSRO but not how many there are or their exact locations.

For each OSRO, we found the shortest available route to the priority site. We added the time to drive to a boat launch (according to Bing maps under "normal" road conditions) and the time to pilot a boat from the launch to the priority site (using an estimated boat speed of five knots).[111] This is very much the best case ETA. In practice, extreme weather leading to road shutdowns, holiday traffic, and other road conditions can influence the ETA. Also, because of data limitations, we could not accurately handle things like the time it takes to receive notification about the spill, to load the boom into a truck or boat, or to set the boom up once it arrives. Without enough information to give an accurate estimate, these are left out of the ETA, a limitation of this analysis. Given these limiting assumptions, if an OSRO could reach the priority site within 12 hours, we counted the length of their boom as available.

Figure 4.1. Approximate Locations of All Oil Spill Response Organizations

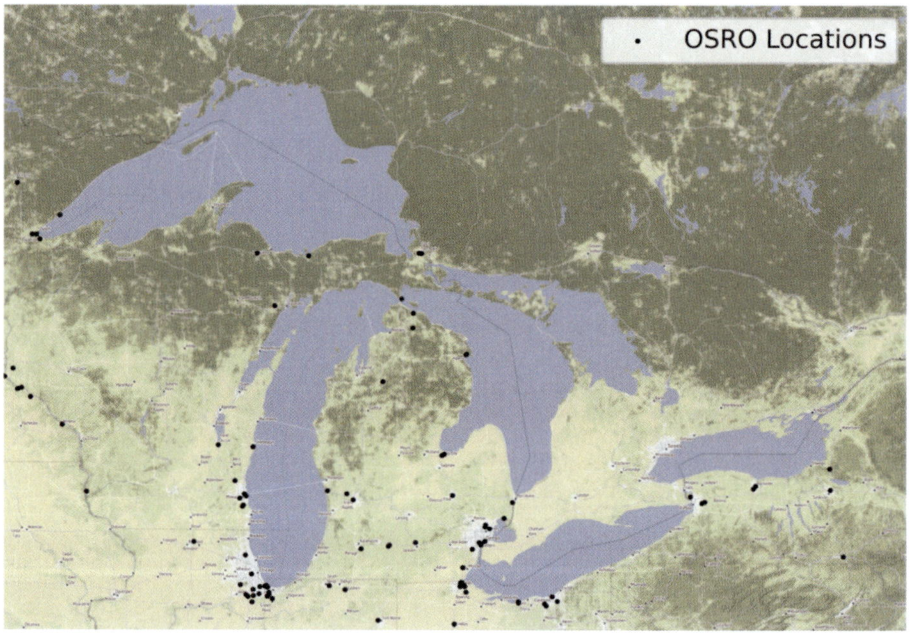

SOURCE: USCG, United States Coast Guard Response Resource Inventory PAV Report, spreadsheet provided to authors, undated c.

[111] Title 33, Chapter I, Subchapter O, Part 155, Subpart D, Section 155.1050, Response Plan Development and Evaluation Criteria for Vessels Carrying Groups I through IV Petroleum Oil as a Primary Cargo.

Figure 4.2. Approximate Oil Spill Response Organization Locations Overlaid with Historic Spills

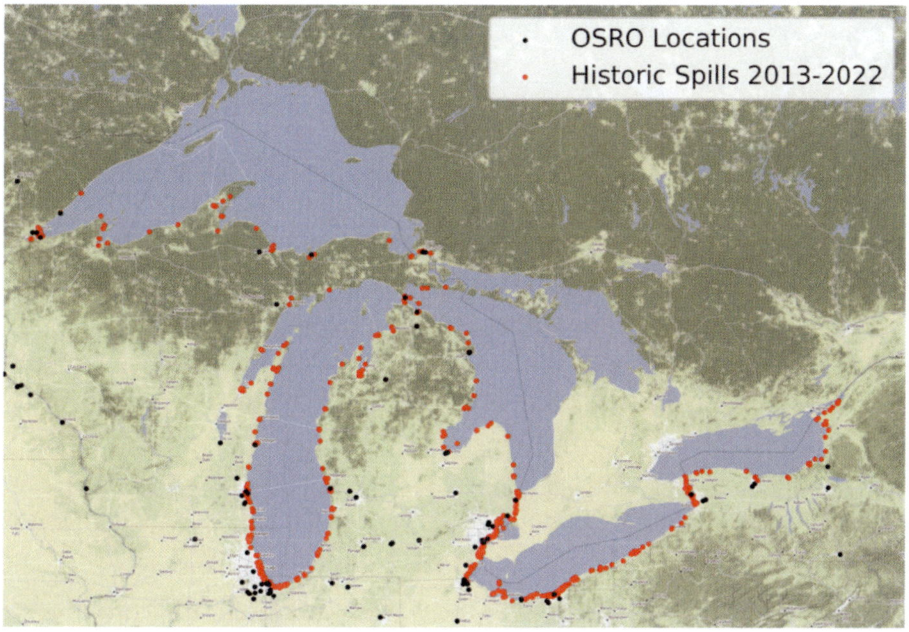

SOURCE: EPA, undated.

Figure 4.3. Heatmap of Risk from Possible Oil Spill Sources Overlaid with Oil Spill Response Organization Locations

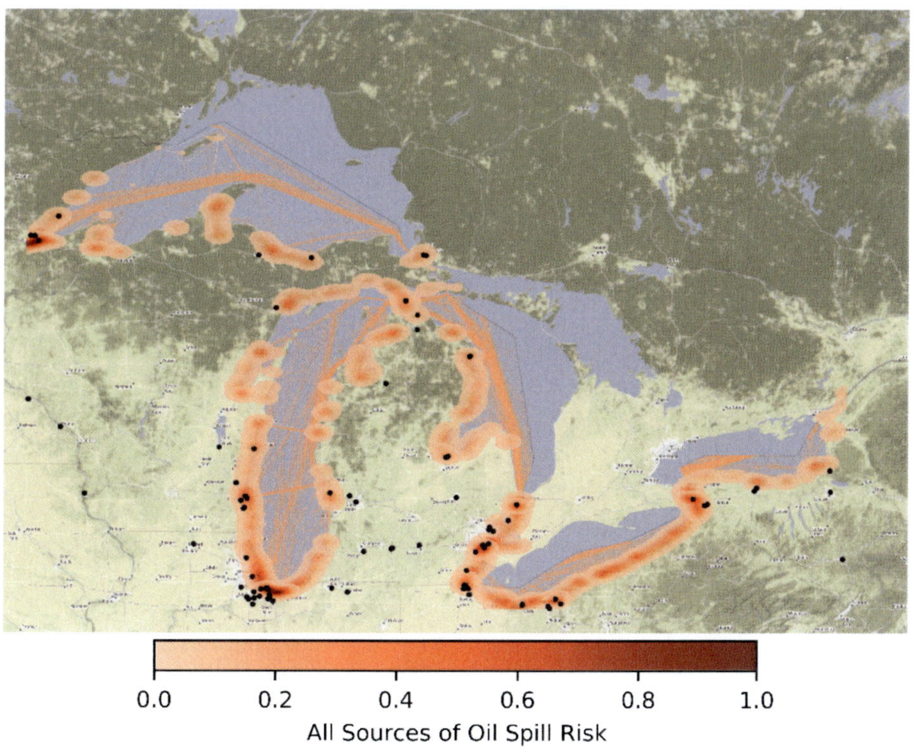

SOURCE: Features information from U.S. Energy Information Administration, undated.

An example priority site is shown in Table 4.1. If an OSRO is able to reach the priority site within 12 hours, their length of boom is added to the available boom. The *minimum* boom is taken from the sector response plans. Each priority site lists an amount of exclusion boom required to keep that site safe.

Table 4.1. Example of Available and Minimum Boom for One Priority Site

Name	Address	Priority	Available Boom	Minimum Boom	Factor of Safety
Siskiwit Bay	E Romans Point Rd Herbster, WI 54844	High	568,535 feet	10,560 feet	53.84

Again, this is only the amount of boom needed to section off a priority site, not the amount needed to clean up the spill itself. But **the available boom from all OSROs that can reach the site is orders of magnitude more than what is required to keep oil from getting in the harbor, and no priority site had insufficient available boom** to do this. Table 4.1 shows the priority site that has the highest amount of minimum required boom, and there is still 50 times more boom available from the OSROs.[112]

These priority sites are on or close to the coast, like a harbor or river mouth, so most of the travel distance is by land. Trucks moving boom to a boat launch are noticeably faster than boats moving boom to the spill, so this only shows that spills near the coast have enough coverage. To handle spills far from the coast, we also looked at how much boom could reach each point in the lake by boat.

For spills in the middle of the lake, the required time of arrival is different. According to 33 C.F.R. 155.1050(d)(1),[113] OSROs must arrive on site with:

> Containment boom in a quantity equal to twice the length of the largest vessel involved in the transfer and capable of being deployed at the site of oil transfer operations—
>
> (A) Within 1 hour of detection of a spill, when the transfer is conducted between 0 and 12 miles from the nearest shoreline; or

---

[112] Authors were not able to verify that the aggregate available boom listed in the various contingency plans are truly reflective of the actual boom that would be physically accessible at the time of an incident. Based upon subject-matter expert discussion regarding historical spill experience, this is an assumption and should be confirmed by response planners on a recurring basis to ensure accuracy.

[113] Title 33, Chapter I, Subchapter O, Part 155, Subpart D, Section 155.1050, Response Plan Development and Evaluation Criteria for Vessels Carrying Groups I through IV Petroleum Oil as a Primary Cargo.

> (B) Within 1 hour plus travel time from the nearest shoreline, based on an on-water speed of 5 knots, when the transfer is conducted over 12 miles up to 200 miles from the nearest shoreline.

The required time of arrival described in paragraph B takes travel time into account when the spill is far from the coast, stating "1 hour plus travel time." Everything other than travel would have to take place in that one hour. Unfortunately, there is no data to analyze how long it takes to detect the spill, coordinate response, get equipment out of storage and loaded onto a truck, etc. Because of that, we cannot verify whether the OSROs can reach a spill in paragraph B's required amount of time. We instead use the 12-hour response time from 33 C.F.R. 154.1045(f) and the five knots on-water speed from 33 C.F.R. 155.1050(d)(1).

The furthest points from shore will take the longest for the OSROs to reach with boom. Figure 4.4 shows a heatmap of available boom on the water, using the estimated five knot boat speed from federal regulation. The black dots represent OSRO locations. The yellow circle around each dot is where these OSROs can reach via boat in the required 12 hours, using a five knot boat speed. The darker the yellow on the heatmap, the more total length of boom can reach that area within 12 hours.

These areas of available boom are very dependent on boat speed, which is not available in the OSRO data. That is why we used an estimate from federal regulation instead. When the boom can be stored on a reel rather than being towed, the boat can go farther faster, so the areas with available boom would increase dramatically. Figure 4.5 shows a map of OSRO lake coverage assuming a faster boat speed of 10 knots. A 12-hour response time at that speed is sufficient to reach almost everywhere with some boom, except for remote areas of Lake Superior. The difference between the coverage of these two maps shows the importance of boat speed in responding to mid-lake vessel spills.

Figure 4.4. Available Boom on Water in 12 Hours, Assuming Five Knot Boat Speed

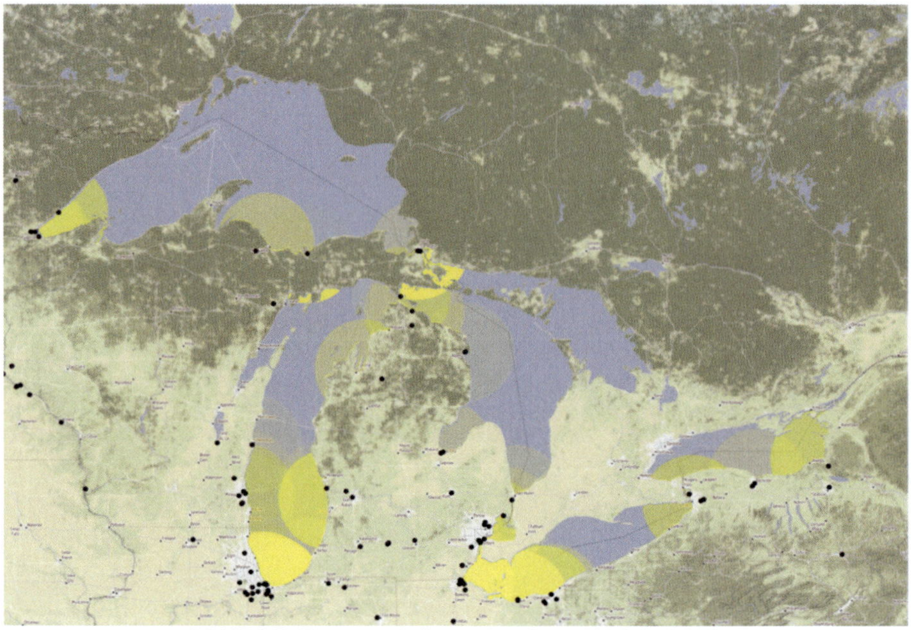

SOURCE: Features information from USCG, 2022.

Figure 4.5. Available Boom on Water in 12 Hours, Assuming 10 Knot Boat Speed

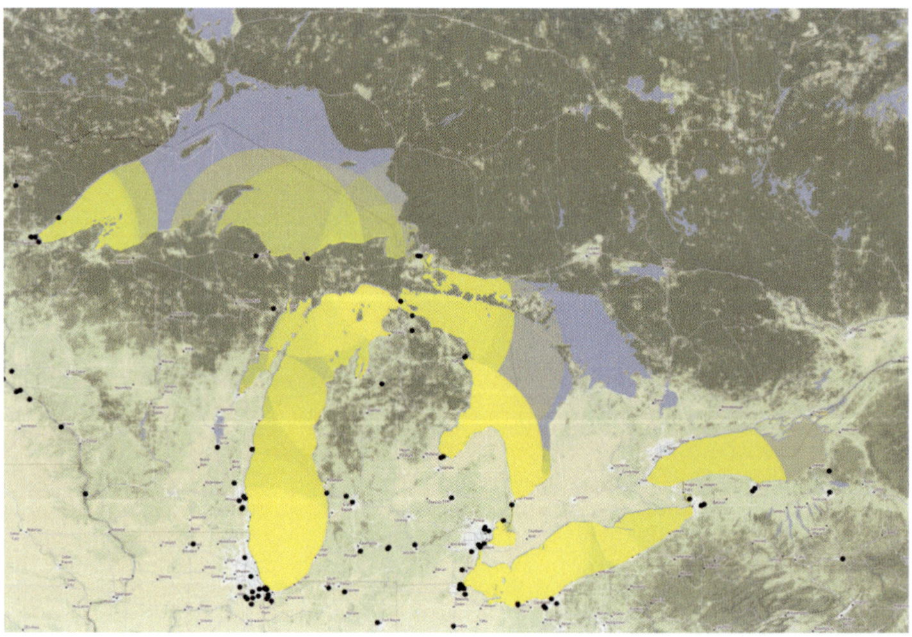

SOURCE: Features information from USCG, 2022

In conclusion, the analysis, based upon the data provided suggests that more than enough boom is available to keep oil from sensitive areas. However, authors recommend that response planners verify on a recurring basis the actual physical location availability and boom condition are accurately reflected in the various response plans. Lake Superior has the least coverage that OSROs can respond to quickly. There are no OSROs near Lake Superior's Keweenaw Peninsula, projecting into the center of the lake, making it difficult to respond to spills in the region in time. There are several tribal jurisdictions on or near the peninsula, which might help to explain the lack of OSROs but adds an additional concern when spills cannot be responded to in time. Spills close to any lake's coast take much less time for OSROs to reach than mid-lake spills. Whether OSROs can reach mid-lake spills in the most remote areas of the Great Lakes within the required response time depends largely on the speed that their boats can get there with boom. It is not uncommon for OSROs to have prepositioned boom where practicable, saving on transportation delays and costs, but in practice this approach might be inefficient if the prepositioned boom is not the correct type or length, or if the condition has deteriorated because of environmental exposure on site.

Chapter 5. Developing Scenarios to Improve Oil-Spill Response

As discussed in earlier chapters, oil-spill responders have a variety of techniques at their disposal, but they have limited opportunities to use them in real-life situations. Many organizations use scenarios to conduct exercises, drills, or tabletop games to give responders practice and practical experience in an operation, to identify any problems or flaws with a plan, and to adjust plans as needed. An organization with robust scenarios will have considered and experienced multiple dimensions that might make a problem easier or harder to approach; using a flatter scenario with fewer dimensions reduces preparedness and decreases the chance of success when responding to a real-life event.

In the case of oil spill scenarios, the nature of oil spills and the variability of environmental factors are great enough that changing just one dimension of a scenario will drastically change the response approach and response requirements. For example, a scenario spill of the same location, volume, and oil type would require a much different response in July than in January when conditions are icy. For this reason, relying on scenarios without a broader framework for developing them can leave an organization unprepared.

For this task, we developed a framework methodology to give the USCG tools to organize and develop scenarios that are robust and that cut across a wide variety of possible spill scenarios. Using a framework to develop scenarios is an organizational method that allows for multidimensional scenario development and also supports the USCG in creating an organizational system to ensure it has covered all the bases when considering oil spill risks.

Identifying Dimensions of Interest

To develop the scenario framework, we began by identifying dimensions of interest that can be used in scenario development. We started with the dimensions identified in already-developed scenarios shared with us by the USCG via ACPs and after-action reports from exercises. We supplemented these dimensions with data identified in NOAA's ERMA Great Lakes mapping application and in USCG Geographic Response Strategies (GRSs), and with the researchers' own subject-matter expertise.

The dimensions were then bracketed and bucketed into factors based on USCG definitions, federal language, and historical environmental conditions. The bracketed buckets and the full framework are available in Appendix B of this document in a template designed to guide scenario development.

Some dimensions are geographically dependent (such as the presence or absence of a potable water intake pipe, or the presence of a rail line adjacent to the Great Lakes), and others are geographically independent (such as most weather conditions). It is important to use data-informed geographic conditions when developing a scenario to ensure that it is useful. Maps and supporting data are also provided and referenced for use in scenario development. Additionally, some dimensions are required for scenario development (e.g., the volume and type of oil), while some dimensions are optional (e.g., the presence of an endangered species at a spill site). The dimensions, geographic dependencies, whether the dimension is required or optional, and sources of the dimensions are shown in Table 5.1.

Table 5.1. Dimensions for Scenario Development

Category of Dimension	Dimensions	Geographically Linked?	Required or Optional?	Source of Dimension
Oil Outflow Information	Source	Yes	Required	USCG scenarios
	Volume	No	Required	USCG scenarios
	Product type	Yes	Required	USCG scenarios
	Potential for additional discharge	No	Optional	Subject-matter expertise
Coordination Information	Responsible party identified	No	Optional	Subject-matter expertise
	Coordination with Canada required	Yes	Optional	Subject-matter expertise
	Spill impacts Tribal lands	Yes	Optional	ERMA
Weather	Beaufort scale (wind and waves)	No	Required	Subject-matter expertise
	Thunderstorms	No	Optional	Subject-matter expertise
	Tornadoes/waterspouts	No	Optional	Subject-matter expertise
	Visibility concerns	No	Optional	Subject-matter expertise
	Ice concentration	Yes	Required	Subject-matter expertise
	Estimate maximum thickness of level ice	Yes	Required	Subject-matter expertise
	Water levels	No	Required	GRS
Additional Risks	Water intake pipes in region	Yes	Optional	GRS
	Presence of/habitat for federally threatened or endangered species	Yes	Optional	ERMA/GRS
	Historic or archaeological site in region	Yes	Optional	ERMA/GRS
Coastal Environment	Shoreline habitat rankings (EIS categories)	Yes	Required	ERMA/GRS

	Managed area	Yes	Optional	ERMA/GRS
Response Health and Safety Concerns	Limitations to workers' physical ability to respond	No	Optional	Subject-matter expertise
	Injuries present	No	Optional	Subject-matter expertise
	Fire or fire risk present	No	Optional	Subject-matter expertise
	Terrorism or criminal factors	No	Optional	Subject-matter expertise
Additional Limitations	Roadways/waterways impeded	No	Optional	Subject-matter expertise
	Holiday weekend	No	Optional	Subject-matter expertise
	Other spill event underway elsewhere	No	Optional	Subject-matter expertise
	Other mission surge event	No	Optional	Subject-matter expertise

Assessment of Comprehensive Scenario Development

Although scenarios themselves need to be multidimensional to reflect the challenges of reality, the other part of scenario development is ensuring that the suite of scenarios comprehensively cover a variety of situations. Practicing for a vessel spill in all types of weather and ice thickness still means that the organization is unprepared for a rail spill. To effectively use this framework, scenario dimensions should be tracked as they are created to make sure that dimensions are not ignored or forgotten.

As an example of how to track scenario dimensions, we took the scenarios and real-life after-action reports that were provided by the USCG (12 in all) and mapped them across the dimensions used in the scenario framework. This analysis is not comprehensive, because we were provided scenarios on an ad-hoc basis.

We mapped the after-action reports across the 27 dimensions we have identified for use in our framework (the full analysis is available in Appendix B). Table 5.2 is a simplified version of what can be developed for future use. The goal of this mapping is to create a visual tracker that clearly displays which dimensions have been thoroughly considered and which dimensions have not been practiced as much. In the example developed in Table 5.2, it is easy to quickly determine that exercises using vessel-based incidents have been well practiced, while exercises considering rail- or facility-based incidents have been less practiced.

Table 5.2. Example of Mapping Past Exercises Against Oil Outflow Information Dimensions

Category of Dimension	Dimensions	Levers/Options	Argo AAR	CANUSLAK 2021 AAR	Manistique ICS 201 and Sector SSM PREP FSE 2018 AAR	Straits Exercise AAR	NOSMAC PREP Full-Scale Exercise AAR	Tecumseh Ground Vessel 2022 AAR
Oil outflow information	Source	Vessel-tanker	Yes	No	No	No	No	No
		Vessel-non-tanker	No	Yes	No	No	Yes	Yes
		Rail	No	No	Yes	No	No	No
		Truck	No	No	No	No	No	No
		Pipeline	No	No	No	Yes	No	No
		Facility	No	No	No	No	No	No
	Volume	Major discharge (greater than 10,000 gallons)	Yes	Yes	No	Yes	Yes	No
		Medium discharge (1,000–10,000 gallons)	No	No	Yes	No	No	No
		Minor discharge (less than 1,000 gallons)	No	No	No	No	No	No
	Product type	Group II: Specific gravity less than .85	No	Yes	No	No	No	No
		Group III: Specific gravity between .85 and less than .95	Yes	No	Yes	Yes	No	No
		Group IV: Specific gravity between .95 to and including 1.0	No	No	No	No	Yes	No
		Group V: Specific gravity greater than 1.0	No	No	No	No	No	No

SOURCE: Features information from Canadian Coast Guard Environmental Response, *CANUSLAK 2021 Exercise After Action Report 18 August 2021*, provided to authors, September 22, 2021b; Canadian Coast Guard Environmental Response, *Great Lakes Joint Response Team Tecumseh Grounded Vessel After Action Report: 18 June–22 June 2022*, provided to authors, September 21, 2022c; LCDR Christopher Pace, USCG, "Scenarios and ACPs? Straits exercise," email to authors, February 7, 2023; USCG, undated a; USCG, *NOSMAC PREP FSE 2022 After Action Report: 05 Apr 2022–17 May 2022*, provided to authors, undated c; USCG, Acquisition Directorate Research and Development Center, *Great Lakes Oil-In-Ice Demonstration 3: Final Report*, CG-D-8-13, revised October 2013; USCG, Manistique incident briefing ICS 201, provided to authors, September 20, 2018b; USCG, *Sector SSM PREP FSE 2018 After Action Report: 19 Sep 2018–20 Sep 2018*, provided to authors, February 15, 2019a.
NOTE: PREP = Preparedness for Response Exercise Program.

Methodology for Scenario Development

The framework for scenario development is meant to be a guiding tool so that different sectors or geographic areas can choose to include or ignore the dimensions that apply to their specific regions or concerns. The framework discussed below is a baseline; geographically specific factors for each sector are discussed in Appendix B.

When using this tool, it is important to be thoughtful and logical. For example, if your scenario has a Beaufort scale of 0–1 (calm, good working conditions), it should not include sleet/hail and the associated visibility concerns. To support the use of data to inform scenario development, users have a few resources at their disposal: transit risk maps, maps of data found in ERMA, and GRS planning documents.

The steps to develop a scenario are as follows:

1. review mapped suite of recently developed scenarios to identify any gaps or less-exercised dimensions
2. identify the region in which scenario will occur
3. reference maps to identify geographically appropriate factors for the region
4. use the framework to choose nongeographically linked factors, considering gaps and less-exercised dimensions and other priorities
5. review chosen factors for logic and alignment with scenario priorities
6. develop a narrative around chosen factors
7. map scenario dimensions onto the suite of recently developed scenarios
8. repeat.

An Example of How to Use the Steps to Complete a Fully Developed Scenario

In Appendix B, we walk through each step to illustrate how they work, culminating in a fully developed scenario. The appendix includes several resources that should be used during scenario development: a framework template, a series of maps, and data to inform current states of oil flow and risk in the Great Lakes region. Additional resources include ACPs and GRSs (GRS products specifically can be useful for scenario development because they have detailed information on a local scale). The ERMA maps referenced in this report are an additional resource rich in data that can support scenario development.

Chapter 6. Oil-Spill Response Challenges

This chapter contains the organizational, operational, and technological challenges for responding to oil spills in the Great Lakes region. The DHS definition of a capability is "the means to accomplish a mission, function, or objective." **We did not identify any remarkable capability gaps,**[114] but gathered examples of where response stakeholders (i.e., the USCG and CCG, responsible parties and their OSROs, and other federal agencies) might not have the full variety of experiences or capabilities needed to respond to an oil spill in an effective or timely manner. Although all the challenges identified could negatively affect oil-spill response, assessing the risks or consequences of an oil-spill response is beyond our scope. In addition, we did not systematically identify challenges where there are plans or research underway to address them.

Challenges to Responding to Oil Spills

In this section we discuss capability challenges that are both contextual and systemic that were gathered and synthesized from our document review, literature search, and structured interviews with stakeholders.

Situational/Contextual Gaps

Weather, environmental conditions, and geography can create challenges for oil-spill response. These challenges are not always an issue, but rather are location- and time-dependent.

Cold Weather Conditions and Freshwater Ice

Ice cover on the Great Lakes varies significantly from year to year and can occur in several forms (such as *fast ice* [fixed ice found along the coastline] or *brash ice* [fragments of ice from other formations that have frozen together]) that affect the methods used for response and recovery operations.[115] Generally, ice is present from early December to late May.[116] A report from 2015 stated that oil-spill response in cold weather and icy conditions is very challenging. At that time, there was experience with ice and cold weather operations in the Arctic that was

[114] For reference, a *capability gap* occurs when "a capability that is required by DHS and/or its stakeholders or partners to perform a mission, function or objective" does not currently exist and there are no plans to provide it in existing programs (DHS, 2016, p. 3).

[115] NOAA, Great Lakes Environmental Research Laboratory, "Ice Cover," brochure, undated b; Michigan Tech, "Definitions of Lake Ice Terms," webpage, undated.

[116] Wang et al., 2017.

supplemented by some governmental analysis, but these approaches and equipment had not been operationally tested in the Great Lakes.[117]

The presence of ice was a concern for several stakeholders. Ice presents a challenge to oil-spill detection, containment, and recovery; vessel movement; and personnel safety. Knowledge gaps identified in our interviews included a lack of both experience with spills in ice and with data or models to understand oil spill behavior and trajectories in ice conditions. For example, oil transport in icy conditions currently is not available in the operational forecast system, but modeling activities are under development. However, note that model development is based on test data of different ice conditions including temperatures and flow rates.[118]

Both Canadian and U.S. sources noted that there are techniques for oil containment and recovery in Great Lakes ice conditions. These approaches can require training in ice profiling and delineation, clearing and marking techniques, operating response equipment, and implementing strategies.[119] Some interviewees had experience recovering oil by creating open water channels for oil skimmers, an approach that is more effective when the oil is concentrated. Other examples come from ECRC, which uses ice slotting, bubbling systems, water injection, and ice removal when responding to spills in ice. SMEs from ECRC provided us with photos, shown in Figure 6.1, that depict some of these techniques.

However, ice does present some challenges. Brash ice is the most challenging to deal with, particularly because bubblers must be used to create open water needed to isolate or concentrate the oil. Participants reported that this works well with smaller spills, but these techniques have not been tested on larger spills (e.g., a 10,000-ton spill) and scaling them up could be problematic. Additionally, oil under ice is particularly challenging to deal with.[120] In general, cold temperatures increase oil viscosity, which makes it more difficult to pump for several reasons. First, pumps must have more power to provide the suction needed, especially when pieces of ice or slush are present or when thicker concentrations of oil occur because of lower evaporation or emulsification rates. Furthermore, hoses, fittings, valves, and pumps might fail when temperatures fall below 32 degrees.[121]

Although these techniques for containing and recovering oil in ice conditions exist (at least since approximately 2015) and are discussed at conferences,[122] our interview

---

[117] Great Lakes Commission, 2015, p. 20.

[118] Interview with NOAA personnel, March 22, 2023.

[119] Briefing provided by response organization, April 15, 2023.

[120] Interview with response organizations, April 21, 2023, and April 25, 2023.

[121] Arctic Council, Emergency Prevention, Preparedness and Response, "Guide to Oil Spill Response in Snow and Ice Conditions in the Arctic," 2015; National Academies of Sciences, Engineering, and Medicine, "Responding to Oil Spills in the U.S. Arctic Marine Environment," National Academies Press, 2014.

[122] Interview with response organization, April 21, 2023.

participants suggest that awareness and experience with them does not appear to be widely shared or generally known by all stakeholders in the region.

Figure 6.1. Canadian Oil Spill Response Organization Ice Response Techniques

ECRC ice response techniques (clockwise from top left): Bubbler system, ice cutter sled; water injection; ice slotting; ice removal

SOURCE: ECRC, Cold Weather Training Document, provided to authors, 2023.

Furthermore, weather conditions can be extremely cold, and personnel safety is a key concern. Particular attention is given to work-rest cycles, heating stations, and appropriate personal protective equipment; monitoring and operations might not proceed if conditions become unmanageable.[123] For example, one USCG oil-in-ice demonstration exercise identified a

---

[123] Interview with U.S. response organization, April 25, 2023.

need for additional personal protective equipment, railings and heated shelters, and an environmentally preferable de-icing solution for equipment.[124] Prolonged exposure of personnel to cold weather conditions and rescue and medical care should a member fall overboard in ice conditions were other personnel health and safety issues that were identified.[125] In general, "cold weather is harder than warm weather response for a multitude of reasons. It's miserable, roads are icy, ice gets in the way of response, equipment breaks down more."[126]

Finally, the USCG is responsible for maintaining open waterways during the normal shipping season and for assisting disabled or stranded vessels in ice-covered waters.[127] The timely availability of ice breakers to free boats from frozen moorings or tankers operating in ice conditions was mentioned as a challenge but came up less frequently as a potential capability gap during our interviews and document review.[128] The USCG notes that although icebreaking vessels can be helpful under certain conditions, they normally are not necessary for a response operation and commercial ice-class tugs are available to augment USCG capabilities if needed.[129]

Other Weather/Operational Conditions, Time of Day

Weather conditions such as high waves, cold temperatures, heavy snow, freezing rain, or ice can delay or complicate a response. For example, wave heights of more than five feet can present unsafe conditions for operating booms and skimmers and might impede some ships from responding.[130] Waterway currents above three knots begin to adversely affect boom operations (and speeds above 10 knots drastically limit their use and efficacy).[131] The trajectory of the oil spill can also be harder to project in fast-moving river systems and require a quicker response to avoid more challenging recovery operations, especially when sensitive areas are nearby and the

---

[124] Products or services that have a lesser or reduced effect on human health and the environment when compared with competing products or services that serve the same purpose. USCG, 2013.

[125] USCG, 2013.

[126] Interview with response organization, April 25, 2023.

[127] Title 14, Section 2, Primary Duties; Executive Order 7521, "Use of Vessels for Ice-Breaking Operations in Channels and Harbors," December 21, 1936.

[128] There is a lot of debate on the proper level of ice breaking capability provided by both large and smaller vessels, particularly to support maritime trade in general. For more on this topic see: Great Lakes Carriers Association, "Great Lakes Icebreaking," position paper, June 16, 2020; USCG, "Great Lakes Icebreaker Acquisition Plan: Fiscal Year 2020 Report to Congress," June 17, 2020.

[129] Coast Guard District 9, "CGD 9 Input for Reply to ELPC Brief," September 18, 2019. The ELPC Brief is "Brief in Support of Plaintiffs' Joint Reply and Opposition to Defendant and Defendant-Intervenor," filed September 13, 2019 and filename "DENVER #673620-v1-elpc_order_sj.docx."

[130] Environmental Law and Policy Center and National Wildlife Federation v. USCG, 447 F. Supp. 3d 602, (E.D. Mich. 2020).

[131] Interview with response organization, April 25, 2023.

spill area rapidly expands.[132] In the Straits, where depths can reach at least 290 feet,[133] oil-spill response could be especially challenging if there is ice cover or fast or unpredictable currents, whereas in Lakes Michigan and Huron, models suggest that oil would disperse rapidly and challenge response operations. In general, "spills that affect the connecting rivers, canals, and channels are more complicated to model, and robust system-wide models have not been developed," according to a Great Lakes Commission report.[134] Additionally, spills that occur near sensitive habitats during mating or migratory seasons can be especially challenging.[135] In cold weather, pumping capacity might also need to exceed the 33 C.F.R. requirement standard. Snow could impede personnel from getting to ships or storage units.[136]

Systemic Challenges

Limited Operational Experience Because of the Low Number of Federalized Spills

Although the relatively low number of spills is clearly good, it does mean that there are few opportunities to gain operational experience or gather information on oil behavior and response efficacy. Furthermore, the low number of spills, that they are generally small, and personnel turnover, all limit how much knowledge transfer occurs.[137] "In the absence of any major oil spills, the capacity, knowledge, and skill set within the [Canadian] government and response organizations have eroded," as one journal article says.[138] These issues are mitigated with training and exercises to some extent. However, previous efforts to practice complex operations have been thwarted by delayed resourcing and a lack of staff with specific experience.[139] For example, there is limited expertise available to responders about wildlife protection, which can be complicated when oiled wildlife cross international boundaries; and managing wildlife issues has not been sufficiently exercised.[140] Canada's OSROs are required to try to scare wildlife away from affected areas, but there are not any strategies for cleaning or rehabilitating animals that become oiled.[141]

Modeling is another challenging area. There are hydrological oil transport models that are used for projecting the trajectory of an oil spill, but there is no formal process to validate them.

---

[132] Interview with Coast Guard sector personnel, March 16, 2023.

[133] D. B. Steinman, Final Geological Report, 1958.

[134] Great Lakes Commission, 2015, pp. 18, 33.

[135] Response organization briefing slides provided April 15, 2023.

[136] Interview with USCG personnel, February 9, 2023.

[137] Interview with response organization, April 21, 2023.

[138] Daisy et al., 2022, p. 19.

[139] USCG, undated c.

[140] Interviews with USCG personnel, February 23, 2023 and March 16, 2023.

[141] Transport Canada, 2013, p. 24.

There are some efforts underway to provide near-real-time imagery of oil transport that could help with this, but real-time data relies heavily on observations from buoys near the coastlines, and models cannot be validated by observational data in some areas of the Great Lakes.[142] Additionally, because of the risks and historical experiences with oil spills, salt water-based research dominates the field. As a result, there are some biases in understanding oil's behavior that might not be true for fresh water; for example, the evaporation rate of diesel fuel is much lower in fresh water.[143]

Coordinating Response Operations Across Agencies

There can be delays or confusion over roles and responsibilities for response operations. For example, although the USCG is the FOSC for spills in bodies of water, other agencies might have the needed technical expertise and equipment to more-readily analyze sediment or soil samples. These agencies might not be able to respond in a timely manner because of resource limits or other factors.[144] Additionally, the USCG's district and sector structure is not well aligned with the other federal agencies' regional offices, which requires additional coordination. The Canadian interviewees also perceived less preparedness and coordination (beyond specific Joint Contingency Plan events) among the responsible Canadian agencies, which can hamper awareness of responsibilities and capabilities.[145] Furthermore, the siloed structure for oil-spill responsibilities among Canadian agencies can hinder comprehensive management. For example, Transport Canada is responsible for certifying OSROs, while the CCG is responsible for on-scene monitoring and coordination. However, since the CCG does not certify OSROs (as the USCG does in the United States), the CCG does not have full awareness of the OSRO capabilities, capacities, and response plans. Lack of awareness of OSRO response planning could hamper response. Additionally, there is no systematic approach to training and exercise for ship-source oil-spill preparedness and response in Canada, and each governmental agency tends to operate its own programs without necessarily involving the full range of stakeholders.[146] That said, there have been instances where each takes responsibility and has capabilities for their respective areas. Furthermore, a Canadian interviewee pointed out that if the CCG does not have authority over response (such as when a shore-side facility, pipeline, or rail car discharges oil into waterways), there could be liability concerns for Canadian and American companies that act on CCG requests, given the language in the Canadian Shipping Act of 2001.[147] These issues

---

[142] A map of the buoy locations can be found here: Great Lakes Observing System, Seagull, web application, undated.

[143] Interview with NOAA, March 22, 2023.

[144] USCG, undated, and Interview with USCG personnel, February 10, 2023.

[145] Interview with Canadian stakeholder, March 9, 2023; interview with USCG sector personnel, March 16, 2023.

[146] Transport Canada, 2013, pp. 19, 24, 32.

[147] Interview with Canadian stakeholder, March 9, 2023, and Transport Canada, 2013, p. 28.

could be concerns in the future because some interviewees cited aging shore infrastructure as an oil-spill vulnerability.

Participants also discussed how some Tribal Nations are more willing than others to collaborate with the USCG. There are 13 tribal entities on 48 reservations that are located within five miles of the coastline,[148] and although participants did not often discuss interactions with tribal nations, they did note that the level of coordination with tribes will vary.[149]

Finally, ensuring that cross-border laws; regulations; sensitivities; and security issues among tribal nations, the United States, and Canada are handled properly is a challenge. Some examples are avoiding equipment getting caught up in customs, declaring chemicals, labor law or immigration violations, or protections for sensitive areas, all of which could delay a response.[150]

Limited Validation of OSRO Capabilities

Some interviewees mentioned that validating OSRO capabilities is limited (audits are the primary mechanism), although it is hard to tell if OSRO capabilities are being counted by multiple locations. However, others noted that when contract support is used, those capabilities (including contract labor) are discounted (i.e., not assessed as 100 percent available) by the USCG.[151] Our geospatial analysis presented in Chapter 4 (Figures 4.4 and 4.5) indicates that priority sites have enough available boom. As described earlier, U.S. response organizations are required to base their response plans on a worst-case scenario, while Canadian response organizations are required to plan for a 10,000-ton spill. Because capabilities for spills above 10,000 tons are not regulated, Transport Canada has no way to know if they are truly prepared for larger spills.[152]

Equipment Positioning and Capabilities

Equipment challenges center on having the right equipment at the right location and having access to technology that enables situational awareness and operational coordination should the need for USCG intervention arise. OSRO and USCG equipment is prepositioned to meet response-time requirements, and interviewees in general did not express concerns about the capacity or accessibility of equipment except in remote areas (e.g., one remote sector noted that it can take 8 to 12 hours to access equipment from the storage location in favorable weather conditions).[153] Another concern for remote areas was the ability to maintain distributed equipment caches stored in trailers throughout the area of responsibility (AOR). In remote

---

148 HSOAC analysis of ERMA (ERMA, accessed June 1, 2023).

149 Interview with USCG Sector personnel, April 17, 2023.

150 Interview with responsible party, March 8, 2023.

151 Interview with U.S. response organization, April 25, 2023.

152 Transport Canada, 2013, p. 14.

153 Interviews with USCG sector personnel, April 3, 2023, and a response organization, April 25, 2023.

AORs, the USCG has supported spill response and might be better equipped than its Canadian counterpart. In addition, because of the lack of boom interoperability between the U.S. and Canada, the CCG capabilities are not force multipliers for USCG (unlike the NSF).[154]

In terms of the types of equipment; both aerial and submersible unmanned systems can help detect, track, and assess a spill; and there is continuing research to improve data acquisition, transfer, processing, and storage and mapping conditions in winter (including with ice) for remotely sensed information.[155] Interviewees noted that the USCG currently has access to just two unmanned aerial systems, one of which cannot operate in temperatures below 30 degrees. In addition, the available submersible must be tethered because it does not have GPS capability, which would be useful to have for ice conditions. Because sharing data off these systems is also cumbersome, they are not capable of providing real-time data. Interviewees felt these systems are very helpful for situational awareness and effective response operations, but the operational constraints and capability limits (not present in newer systems) limit their utility. They also noted issues with training and retraining unmanned vehicle pilots.[156]

Although communications equipment did not come up in the interviews, several after-action reports identified operational communications issues related to bandwidth and connectivity of communications and computer equipment that hindered peer-to-peer communications and USCG to CCG communications. Some examples shared by participants were the limits of communications with hailers and the challenges with using cell phones during cold-weather operations (involving batteries, operating phones while using personal protective equipment, and procedures).[157]

Barriers to Accessing New Technologies for Dealing with Oil Spills in Fresh Water

There are policy, resourcing, stakeholder, and market disincentives to innovate. As mentioned previously, response requirement standards are capability-driven, not performance-driven, and response is limited by law to mechanical means unless exceptional conditions apply.[158] In general, there is reluctance because of stakeholder beliefs around nonmechanical means, even though under certain circumstances they can be both more efficient and relatively safe for the environment. For example, ISB has been determined to remove more than 90 percent

---

[154] Interview with USCG sector personnel, March 16, 2023.

[155] USCG, 2022.

[156] Interviews with USCG personnel February 9, 2023, and April 3, 2023.

[157] USCG, 2022; Canadian Coast Guard Environmental Response, 2022a; Canadian Coast Guard Environmental Response, 2022b; CANUSLAK Joint Response Team, Annual Meeting Minutes, March 14, 2022; USCG, undated a; USCG, 2013.

[158] LimnoTech, 2018; Title 40, Chapter I, Subchapter J, Part 300, Subpart J, Section 300.910, Authorization of Use; Interview with Responsible Party, March 8, 2023.

of spilled oil and can be an effective option in ice conditions or when the oil is concentrated.[159] However, there are potential downsides that must be managed. Most notably is the concern over the health and environmental consequences from the smoke plume that carries pollutants, particularly when combustion is incomplete (there has been some research to address these issues but more study is needed).[160] In contrast, one Canadian study reported that mechanical recovery is not feasible in all weather or environmental conditions and only collects between 5 and 15 percent of the spilled oil even under optimal conditions.[161] There are few incentives to develop better-performing technologies. One issue is that Canada has no legal framework for approving the use of nonmechanical response techniques that might use dispersants, solidifiers, or cleaning agents (this issue is exacerbated by the fact that there is no available data on their use in fresh water).[162] Second, stakeholders have differing priorities for managing oil spills given that the Great Lakes region is a drinking water source, a recreational area, and a home to sensitive habitats for wildlife. Third, robust processes for evaluating the environmental benefits against these concerns do not exist. For example, ISB might offer net environmental benefits in certain conditions, but stakeholder concerns, lack of funding, and legal complexities hinder their use.[163] Finally, the small number of federalized spills in fresh water do not create much demand for new approaches.

Inconsistent Approaches to Assessing Risks and Vulnerabilities

Ways of assessing risks and vulnerabilities in the ACPs and differing levels of integration between the ACPs and the Geographic Response Plans (GRPs) were seen as inconsistent. Participants stated that sectors differed in how much they integrated the two documents; some felt that greater integration would facilitate awareness of at-risk resources and minimize duplication of effort. For example, ACPs were mixed on the inclusion of water intakes, which can vary by type in terms of how the water is drawn and the distance from the shoreline. Participants also shared that drinking water managers are not always promptly contacted even if the spill occurs near their facility.[164] Some saw the ACP as too general to be useful and relied on the GRP instead.

The interview sessions also revealed that there are differences in how risk is assessed. For example, industry is focused on the statute's requirement that provides a definition for worst-

---

[159] Ian Buist, James McCourt, Steve Potter, Sy Ross, and Ken Trudel, "In Situ Burning," *Pure Applied Chemistry*, Vol. 71, No. 1, 1999.

[160] F. Bugra Demirel and Ersan Basar, "Applications of a New Product To Decrease the Smoke Plume Opacity of In-Situ Burning of Oil Spills," *Fresenius Environmental Bulletin*, Vol. 25, No. 11, 2016.

[161] Transport Canada, 2013, p. 9.

[162] Transport Canada, 2013, p. 26 and Interview with District 9 personnel, February 9, 2023

[163] Daisy et al., 2022; Interview with scientist, March 8, 2023; Transport Canada, 2013.

[164] Southeast Michigan Council of Governments member, video interview with the authors, April 28, 2023.

case discharge while the national contingency scenario planning uses a different definition. New risk guidance is being developed. In Canada, responsible parties and OSROs use a national standard of having equipment on-site within 72 hours for a 10,000-ton spill, which does not account for unique regional or area sensitivities, vessel traffic, oil movement, and other sources of risk.[165] There are also unknown risks associated with sunken vessels, shoreside underground storage tanks, and aging infrastructure, especially since the locations, type, and quantity of pollutants, condition, and other factors have not been identified.[166] The fate and trajectory of spills that originate from shipwrecks, underground storage tanks, or leaks from other aging infrastructure cannot be modeled when the product is unknown.[167]

Lack of Information About Certain Types of Oil

Although there are methods for dealing with nonfloating oils, these response operations are complex, and not all response organizations are equipped to handle them. The behavior of nonfloating oils in fresh water is not well-understood because most spills occur in salt water. These oils differ from floating oils in both behavior and environmental effects (particularly on resources in the water column and lakebed). As the Northern Area Contingency Plan states:

> Although floating oil modeling and predictions are well developed, models and predictions of heavy-oil behavior are unverifiable and rarely used. There is a lack of supporting field data due to the complex nature of three-dimensional currents when oil sinks into the water column. Field data can be verified, but methods are very slow and labor intensive that make updating spill models difficult. Remote sensing equipment is very limited in its use because it cannot penetrate the water column.[168]

Furthermore, thicker oils in cold temperatures are more difficult to remove with pumps because of the amount of suction required.[169] In addition, some oils are toxic or flammable and present a health and safety hazard to personnel, and some water intakes are relatively far offshore and deep, so sinking oil could be an issue.[170] A 2015 report also notes, "According to the Coast Guard, adequate response methods and techniques do not currently exist for marine related incidents, e.g., spills of heavy oils to open bodies of fresh water such as the Great Lakes."[171]

---

[165] Transport Canada, 2013; Daisy et al., 2022.

[166] Interviews with federal agency and USCG personnel, NOAA, USCG Buff, USCG MER and Transport Canada, 2013.

[167] Interview with NOAA personnel, March 22, 2023.

[168] USCG, Northern Michigan Area Contingency Plan, Sector Lake Michigan, 2017f.

[169] International Tanker Owners Pollution Federation Limited, *Use of Skimmers in Oil Pollution Response*, Technical Information Paper 5, 2014b.

[170] Interviews with response organization and USCG sector personnel, April 25, 2023, and March 2, 2023, respectively.

[171] Great Lakes Commission, 2015, p. 30.

Conclusion

In conclusion, we did not identify any remarkable capability gaps in response capabilities. However, we did compile several issues that could challenge the effectiveness or timeliness of an oil-spill response in the Great Lakes region.

Some challenges are situational or contextual, involving weather, environmental conditions, and operational conditions that could challenge a response. These challenges are not always an issue, but rather are location- and time-dependent. Additionally, techniques for oil containment and recovery in ice conditions exist, but they do not appear to be widely shared.

Other challenges are systemic, involving limited operational experience, coordination across agencies, and equipment positioning and capabilities. Additional issues are the limited validation of OSRO capabilities, barriers to accessing new technologies, inconsistent approaches to assessing risks and vulnerabilities, and a lack of information about certain types of oil.

Chapter 7. Conclusion

Overall, it appears that the Great Lakes response community has clear roles and responsibilities, robust response capabilities, and risks that are being managed in accordance with existing regulations and policy. Because the community is well positioned to respond to oil spills, we make no recommendations but offer a few observations for consideration.

Better Communication Could Be Beneficial for Stakeholders and the Public

There is a high level of cooperation and willingness to collaborate, but some agencies are siloed and could improve their external communications, information-sharing, and contributions to educating concerned citizens and community leaders about oil-spill response tactics, techniques, and procedures (TTP). After interviewing a cross-section of responders, we found that potential cultural divergence among key stakeholders might be overcome by more collaboration and information-sharing, which should improve as the Great Lakes Oil Spill Center of Expertise grows into its new role as a regional leader, and builds its reputation as the go-to organization for scientific subject-matter expertise, key facilitator for translational research, and a national repository of knowledge.[172] An example of potential cultural divergence is the perception that the USCG naturally defaults to coastal salt water environment response TTP, which surfaces in minor but potentially significant ways: overemphasizing vessel spills in exercises, excluding drinking water contacts from emergency call lists, and focusing on response time requirements and mechanical approaches as opposed to the efficacy of the method.

Additional Scenarios Could Reveal Gaps in Capabilities or Capacities

Although the Great Lakes response community appears to have clear roles and responsibilities, there are gaps in what the existing scenarios cover. Developing scenarios more systematically and collaboratively could uncover gaps in both capability and capacity. To ensure there are not any capability gaps, in addition to better incorporating stakeholder concerns, planning scenarios should also be anticipatory and consider future conditions resulting from climate change, regional development patterns, or changes in the oil transport modes and volume.

---

[172] Anna Jean Wirth, Dulani Woods, Katherine Anania, Gary Cecchine, and Debra Knopman, *Developing the Great Lakes National Center of Expertise for Oil Spill Preparedness and Response: An Opportunity to Reduce Risk and Impacts of Future Spills in Freshwater*, Homeland Security Operational Analysis Center operated by the RAND Corporation, RR-A1222-1, 2021.

Appendix A. Detail on Capabilities of Oil-Spill Responders

This appendix to Chapter 3 provides more details about the capabilities of oil-spill responders.

Emerging Research on Oil-Spill Response

Although there is a general lack of fresh water oil spill research currently available, Canada is leading the way in expanding this research, specifically in the International Institute for Sustainable Development Experimental Lakes Area (IISD-ELA) program. The IISD-ELA, located in Northwestern Ontario, includes a "natural laboratory comprised of 58 small lakes and their watersheds set aside for scientific research."[173] The lakes are isolated and provide a valuable area where controlled oil spills can be conducted to study the effects of oil on fresh water ecosystems in areas unaffected by human activities. Among other things, research in the IISD-ELA has shown how diluted bitumen acts in fresh water,[174] how surface oil effects macroinvertebrates,[175] and how the concentration of polycyclic aromatic compounds change over time in fresh water lakes.[176] Further research in the IISD-ELA could provide additional insight on the interaction between oil and fresh water ecosystems and be a testing area for such emerging oil-spill response techniques as chemical dispersants and ISB.

Monitoring

An example of monitoring equipment employed in the Great Lakes region can be found in Michigan. The Southeast Michigan Council of Governments, a regional planning organization, secured funding for and implemented a water monitoring program upstream from regional water treatment facilities. State law requires testing to be conducted *after* a facility treats the water, but

---

[173] International Institute for Sustainable Development, Experimental Lakes Area, "Who We Are," webpage, undated.

[174] Jose Luis Rodriguez-Gil, Sawyer Stoyanovich, Mark L. Hanson, Bruce Hollebone, Diane M. Orihel, Vince Palace, Robert Faragher, Fatemeh S. Mirnaghi, Keval Shah, Zeyu Yang, Tyler A. Black, Jeffrey Cederwall, Johanna Mason, Samuel Patterson, Lauren Timlick, Jonathan Y. Séguin, and Jules M. Blais, "Simulating Diluted Bitumen Spills in Boreal Lake Limnocorrals: Part 1–Experimental Design and Responses of Hydrocarbons, Metals, and Water Quality Parameters," *Science of The Total Environment*, Vol. 790, 2021.

[175] T. A. Black, M. S. White, J. M. Blais, B. Hollebone, D. M. Orihel, V. P. Palace, J. L. Rodriguez-Gil, and M. L. Hanson, "Surface Oil is the Primary Driver of Macroinvertebrate Impacts Following Spills of Diluted Bitumen in Freshwater," *Environmental Pollution*, Vol. 290, 2021.

[176] S. Stoyanovich, Z. Yang, M. Hanson, B. P. Hollebone, D. M. Orihel, V. Palace, J. R. Rodriguez-Gil, F. Mirnaghi, K. Shah, and J. M. Blais, "Fate of Polycyclic Aromatic Compounds from Diluted Bitumen Spilled into Freshwater Limnocorrals," *Science of The Total Environment*, Vol. 819, 2022.

the Southeast Michigan Council of Governments set up upstream monitoring. If an oil spill were to occur, it would be detected before contaminated water made it into the water treatment plant. This upstream monitoring provides a warning to water treatment facilities and would allow them to coordinate water service shutdowns and begin emergency distribution before contaminated water is taken into the facility.[177]

Mechanical Means

Booms

The typical boom has four elements: a freeboard, which is an above water barrier used to contain the oil; a flotation device that allows the boom to stay on the water's surface; a skirt that hangs below the surface and prevents oil from spreading underneath the boom; and a longitudinal support that runs along the skirt and strengthens the boom, preventing too much movement.[178]

Skimmers

The three primary types of skimmers are the weir skimmer,[179] oleophilic skimmer, and the suction skimmer.[180] Each skimmer operates most effectively when thick oil slicks float on the water's surface. Oleophilic skimmers have the best chance of working effectively under such adverse conditions as choppy waves, significant debris, or sinking oil. Weir and suction skimmers can become clogged or inefficient under less than ideal environmental circumstances.

---

[177] Southeast Michigan Council of Governments member, video interview with the authors, April 28, 2023.

[178] EPA, "Booms," October 2022d.

[179] Weir skimmers operate by taking the surface layer off the water. Oil flows into the central hopper where it's pumped to storage (Elastec, "Weir Skimmers," webpage, undated).

[180] A suction oil skimmer works like a household vacuum cleaner on the surface of the water. Its floating head sucks oil into a holding tank (Elastec, undated).

Figure A.1. Types of Weir Skimmers

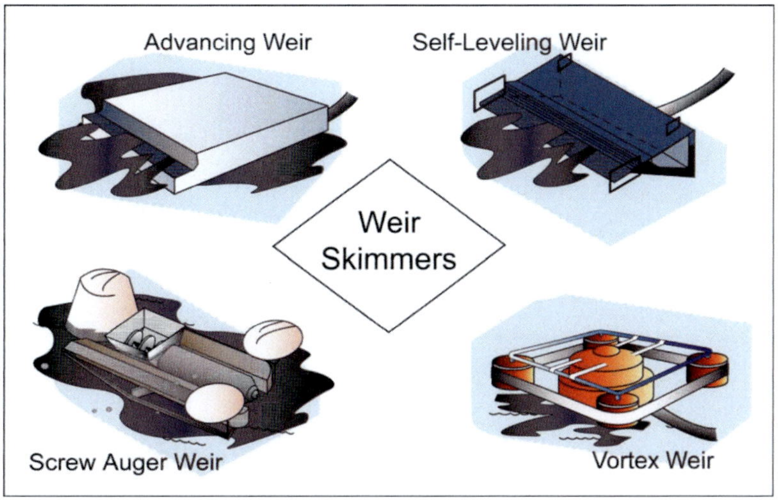

SOURCE: Reproduced from Nuka Research and Planning Group, *Inventory and Assessment of Marine Oil Spill Response Resources in Massachusetts and New England States*, December 2009.

Figure A.2. Types of Oleophilic Skimmers

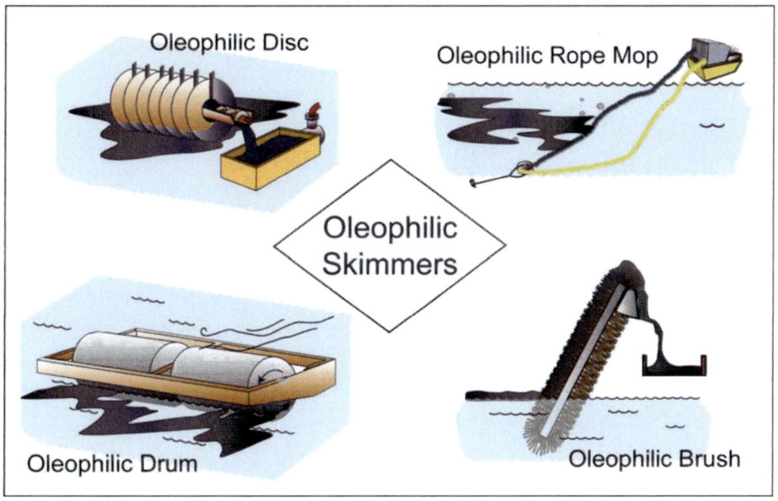

SOURCE: Reproduced from Nuka Research and Planning Group, 2009.

Figure A.3. Types of Suction Skimmers

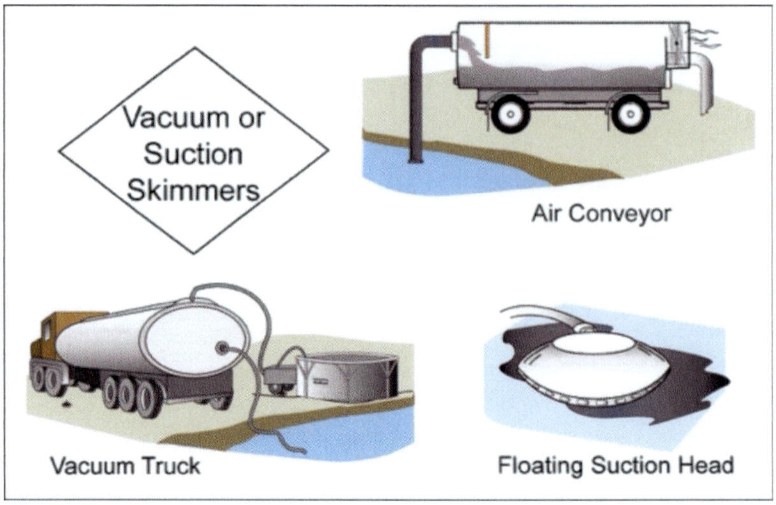

SOURCE: Reproduced from Nuka Research and Planning Group, 2009.

Nonmechanical Means

In-Situ Burning Considerations

Per the ACPs governing the Great Lakes, ISB is authorized for use and "may offer a logistically simple, rapid, inexpensive, and relatively safe means for reducing shoreline impacts of an oil spill."[181] Before an ISB is authorized, the FOSC must consult with and receive approval from the appropriate authority in the state where the response is occurring, and the Department of the Interior.[182] Additionally, Regional Response Team 5 prepared an ISB annex to the RCP that recommends consultation with NOAA, effected tribal governments, and Canadian authorities when necessary, prior to initiating an ISB.[183] According to the NCP, FOSCs are authorized to use burning agents, which are chemicals that create a more effective burn, for an ISB, and the EPA does not require data about these agents to be submitted to the agency and does not include the agents on the NCP product schedule. The FOSC will work with state and federal officials to determine if a burning agent will be used for an ISB.

The USCG conducted a series of test burns to attempt and quantify the burn efficiency (BE) of an ISB in fresh water. The tests found that BE can vary significantly depending on the size of the spill, the slick thickness, and type of oil. BE in this series of experiments ranged from 58–97 percent. The research suggests that a 10mm thick oil slick is likely a reasonable thickness to initiate an ISB without an accelerant. The research showed that for large-scale burns, nearly all

---

[181] USCG, 2015. All other ACPs we reviewed have similar language pertaining to this topic.

[182] USCG, 2015. All other ACPs we reviewed have similar language pertaining to this topic.

[183] Region 5 Regional Response Team, *In-Situ Burning Annex*, undated.

of the oil was burned off. Of the four trials, three had a BE of 97 percent or higher.[184] A separate series of studies conducted in 2018 and 2019 showed that BE can vary significantly depending on the type of oil. BE was estimated to be between 80 and 91 percent for Oseberg crude oil but below 60 percent for two heavier fuels used in the studies.[185]

Chemical Dispersant Considerations

Laboratory studies have found that dispersants are less toxic than the oils they interact with,[186] but less is known about the consequences of chemical dispersants in field environments, and even less is known about the effects in fresh water environments.[187] Studies have also shown that concentrated oil dispersed in the water column can be harmful to aquatic life,[188] but the benefits provided from quickly breaking down an oil spill using chemical dispersants in fresh water has not been quantified against the known harms. It is particularly important to understand how chemical dispersant use could affect the people who live near, operate in, and drink from the Great Lakes. A focused research effort directed at understanding how chemical dispersants impact fresh water environments and the possible implications for nearby populations could advance the use of chemical dispersants as an acceptable response technique. Figure A.5 illustrates three methods of applying chemical dispersants. Dispersants can be applied to the surface by plane (at top left) or from a boat (at top right). Underwater remotely operated vehicles can even apply dispersants closer to the source of the spill (bottom right of figure A.5).

---

[184] USCG, Acquisition Directorate Research and Development Center, *Freshwater In-situ Oil Burning*, CG-D-01-21, February 2021c.

[185] Liv-Guri Faksness, Frode Leirvik, Ingrid C. Taban, Frode Engen, Hans V. Jensen, Jan Willie Holbu, Hilde Dolva, and Magne Bråtveit, "Offshore Field Experiments with In-situ Burning of Oil: Emissions and Burn Efficiency," *Environmental Research*, Vol. 205, 2022.

[186] Adriana C. Bejarano, "Critical Review and Analysis of Aquatic Toxicity Data on Oil Spill Dispersants," *Environmental Toxicology and Chemistry*, Vol. 37, No. 12, 2018.

[187] The French company Cedre tests and approves chemical dispersants specifically for use in fresh water. As of April 2023, the company had two approved chemical dispersants for use in fresh water (Cedre, "Dispersants Approved by Cedre For Oil Spill Response," April 2023). This is contrast to the EPA NCP product schedule which lists 20 approved dispersants but does not differentiate between marine and fresh water use.

[188] LimnoTech, 2018.

Figure A.5. Three Methods for Applying Chemical Dispersants

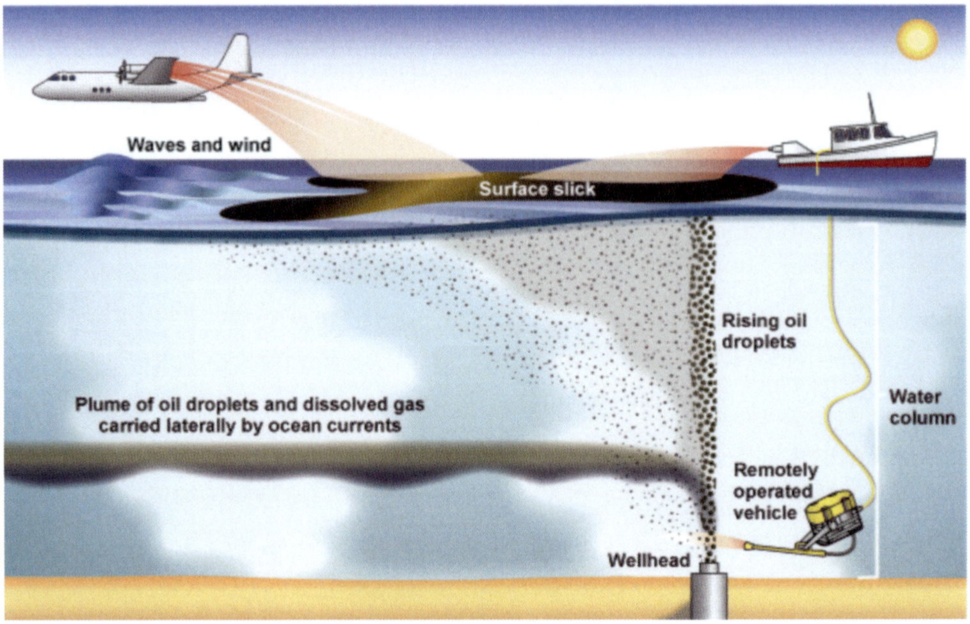

SOURCE: Reproduced from GAO, 2021.

Mechanical Means in Icy Environments

Oil-spill response in ice mainly relies on gaining access to the oil. Under thick sheets of ice, cutting equipment (like the devices pictured in Figure A.6) can be a relatively effective way to gain access. Even partial ice coverage can still significantly impact response capabilities.

Figure A.6. Example of Chainsaw Sled Creating a Space for Oil to Collect

SOURCE: ECRC, 2023.

Specialized Equipment

There are some devices designed specifically to clean oil from ice on the water. Finland has produced equipment such as the Lori Ice Cleaner and LOIS System that operate on the water's surface pulling chunks of ice into the device, cleaning them, and then returning them to the water. These specialized skimmers are pushed over the water's surface, forcing ice chunks beneath the skimmer where they are cleaned, and the oil is collected in the device where it can be recycled or disposed of.

Appendix B. Framework and Data for Scenario Development

This appendix provides instructions and tools for developing scenarios that are used to practice and improve organizational preparedness for responding to oil spills. We provide several types of tools:

Step-by-Step Instructions for Developing Scenarios

The first section has instructions that walk through the eight steps for developing scenarios and point to the tools needed to do so.

Inventory of Previous Scenarios

Table B.1. shows previous scenarios that have already been used for exercises and lists the dimensions used to build them. This inventory shows the developers of new scenarios what dimensions have already been practiced so that they can include less-practiced dimensions in the new scenarios. The inventory should be a living document to which a new column is added for each new scenario and the dimensions it includes.

Template for Developing Scenarios

We created a fill-in-the-blank template that should be used to guide scenario-building. Step-by-step directions for using the template are laid out in Chapter 5, and the full template is shown in figure B.16. To create a useful and easy-to-use framework, each dimension developed for scenario-building is supported by data for the Great Lakes region. These data are bracketed, meaning the realm of the possible lies between the minimum and the maximum values. The USCG already has many of these brackets defined in legislation (such as the worst-case discharge), and we supplemented those definitions with others provided by NOAA and other government organizations when needed. Many of these factors were already categorized into buckets by the USCG, NOAA, or other federal organizations; we used our subject-matter expertise to categorize the remaining buckets. The template includes both the bracketed and categorized factors.

Reference Maps for Geographically Linked Dimensions

This section displays a series of maps that show relevant features of the Great Lakes region. These features (e.g., locations of facilities that might be the source of a spill, water intake pipes, tribal lands) will help developers create new scenarios that are logical and realistic.

Predefined Risks That Are Sector-Specific

The last section of this appendix summarizes pre-defined risks for the different Great Lakes sectors that can be helpful for identifying unique regional concerns and for guiding prioritization for scenario development.

Step-by-Step Instructions for Developing Scenarios

As an example, we will walk through the practice of developing a scenario, which will culminate with a fully developed scenario. There are several resources that should be utilized during scenario development. We have developed a framework template (Figure B.16), a series of maps, and data to inform current states of oil flow and risk in the Great Lakes region. Additional resources include ACPs and Geographic Response Strategies (GRS products specifically can be useful for scenario development, as they have detailed information on a local scale). The ERMA maps referenced in this report are an additional resource rich in data that can support scenario development.

We encourage anyone embarking on scenario development to consult with local and regional experts. The resources we have provided represent the best available data from a snapshot in time. Petroleum product transit, development, environmental conditions, and policies and regulations are all variable and will change over time. To ensure scenarios maintain their relevance, local experts should be involved in developing scenarios and informing updated data.

Step 1. Review Mapped Suite of Recently Developed Scenarios to Identify Any Gaps or Less-Exercised Dimensions

Because this is an example scenario, we will use the mapping of scenarios shown in Table B.1 for this exercise. In practice, Table B.1 will be built as the framework is used over time.

For this example, we want to ensure we have covered our bases with respect to scenario development. After reviewing Table B.1, our limited analysis suggests that a few dimensions have been exercised less recently. Therefore, some dimensions to consider are the following:

- Facility-based spill
- Group V oil
- a higher Beaufort scale
- extreme weather such as thunderstorms, tornadoes, visibility concerns
- lower- or higher-than-average lake depth
- water intake pipes impacted or at risk
- presence of endangered species
- presence of historic or archeological site
- diversity of shoreline habitat types
- injuries present
- terrorism or criminal risk factors

- other limitations such as impeded roadways, holiday weekends, personnel or OSRO shortages.

For our example scenario, we are prioritizing practicing a facility-based spill (an orphan oil well) near a water intake pipe. It should be noted that the purpose of this step is to see how the scenario being developed exercises skills and challenges relative to other recently developed scenarios, exercises, and real-life responses. Leadership and organizers of the scenario might have other priorities for what to practice, in which case this analysis does not need to be the driver of the scenario dimensions.

Steps 2 and 3. Identify the Region Where the Scenario Will Occur, and Reference Maps to Identify Geographically Appropriate Factors for the Region

Identifying a region can be a chicken-and-egg problem, depending on the priorities of the scenario. If we needed to practice in a specific region, we would choose that region as the starting point. But for this example, we are prioritizing a facility-based spill near a water intake pipe from the "source" and "water intake pipes" options shown in Figure B.1 rather than a geographic area, so we first need to reference maps to identify a location with both facilities that could cause an oil spill and a nearby water intake pipe.[189]

Figure B.1. Filling in the Framework Template—Source and Water Intake Pipes

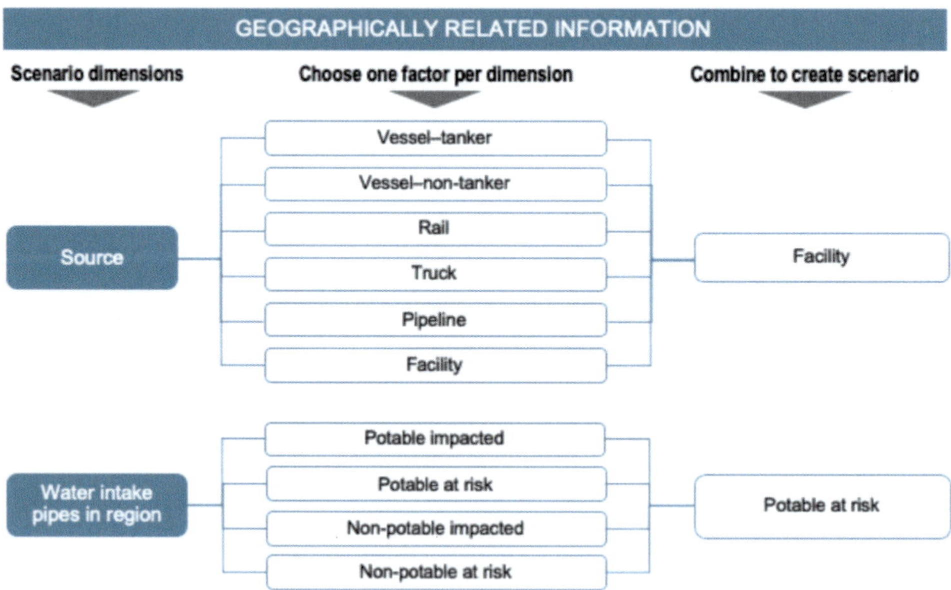

[189] The locations of water intake pipes are not always publicly available for security reasons. Sometimes the ACPs contain water intake pipe information and sometimes that information is scrubbed, depending on the sector or state policies. For this report, we are using a site in Lake Erie because water intake pipe information for Ohio is publicly available in ERMA and is not considered sensitive.

Referencing maps for facilities and water intake pipes, we chose a location east of Cleveland, past Euclid, Ohio, as shown in Figures B.2 and B.3. According to U.S. Energy Information Administration, the facility includes orphan oil wells,[190] and according to ERMA, there is a nearby water intake pipe that supports the Cleveland Public Water system.

Figure B.2. Subset of "Facilities Map"

SOURCE: Features information from U.S. Energy Information Administration, undated.

[190] Ohio has 20,000 known orphan oil wells, and the state is prepared to begin plugging 1,053 of those that are known, some of which are on the shores of Lake Erie (Peter Krouse, "Cuyahoga County Has More Than 300 Orphan Oil and Gas Wells That May Need Plugging," Cleveland.com, March 24, 2022).

Figure B.3. Subset of "Water Intake Map"

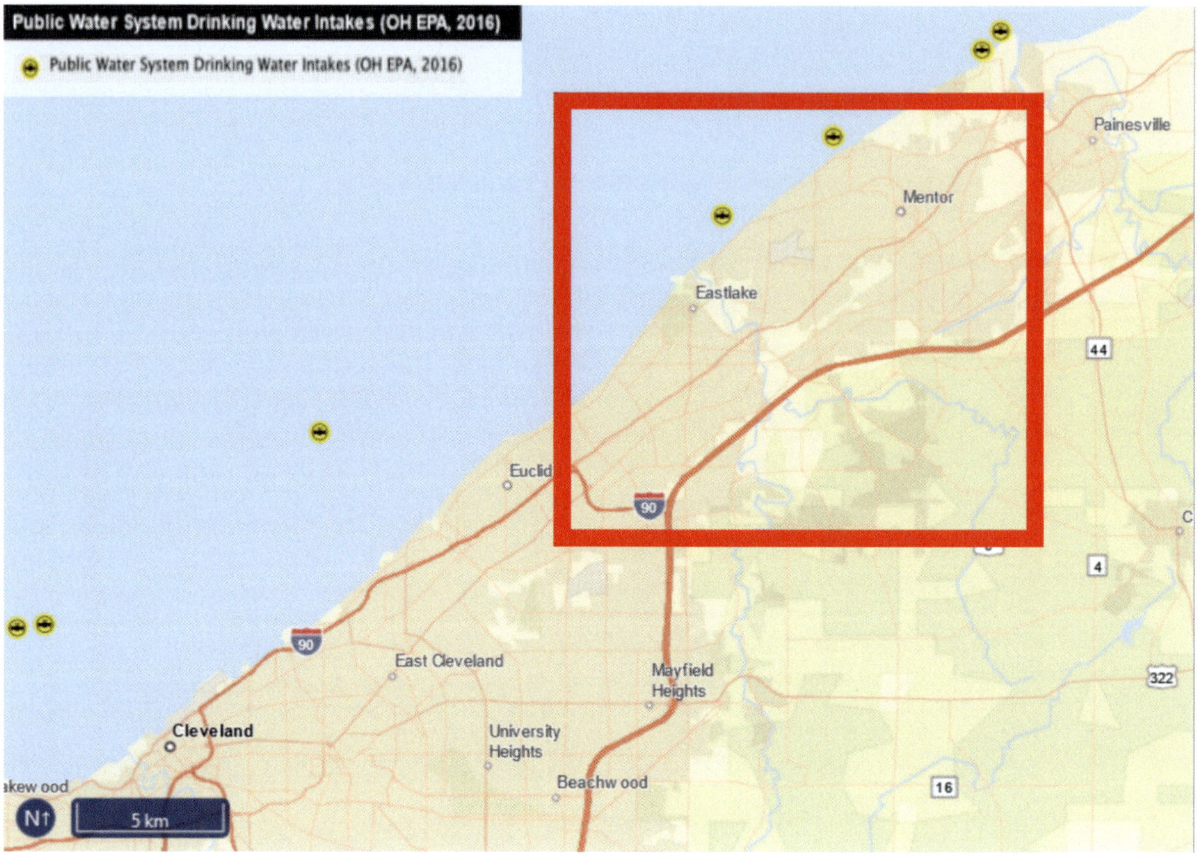

SOURCE: Adapted from ERMA, 2023.

Now that the region has been chosen, the other geographically linked dimensions can be checked against the maps. Using the template as a guide, the next category, "Coordination with Canada required," should be marked "no" because the spill is sufficiently far from the Canadian border, as shown in Figure B.5.

Next, reference the "Tribal Lands Map" in Figure B.4 (see the green shaded area). There is no tribal land immediately east of Cleveland, so the category will not apply to this scenario and should be marked "no" in the framework template, as shown in Figure B.5.

Figure B.4. Subset of "Tribal Lands Map"

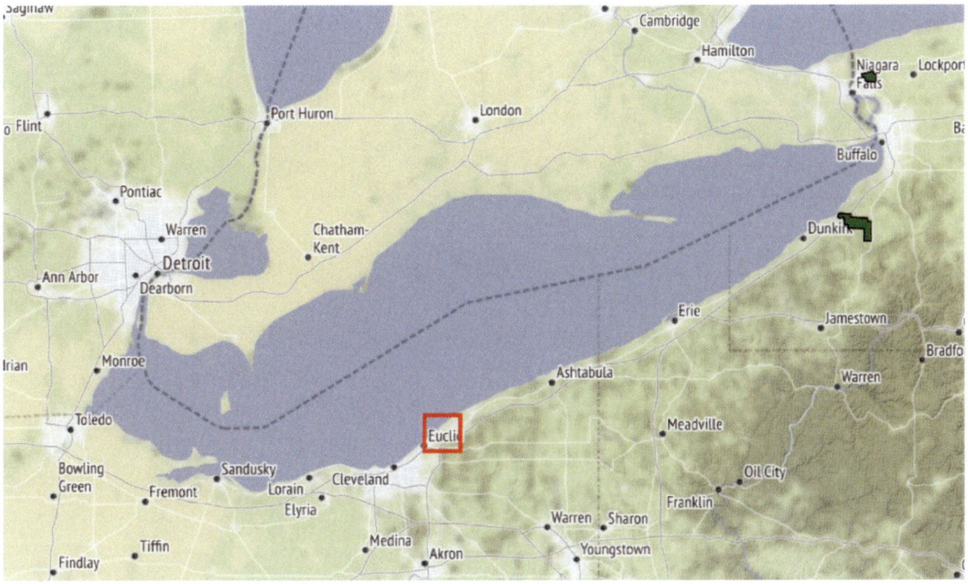

SOURCE: Adapted from ERMA, 2023.

Figure B.5. Filling in the Framework Template—Canadian and Tribal Lands

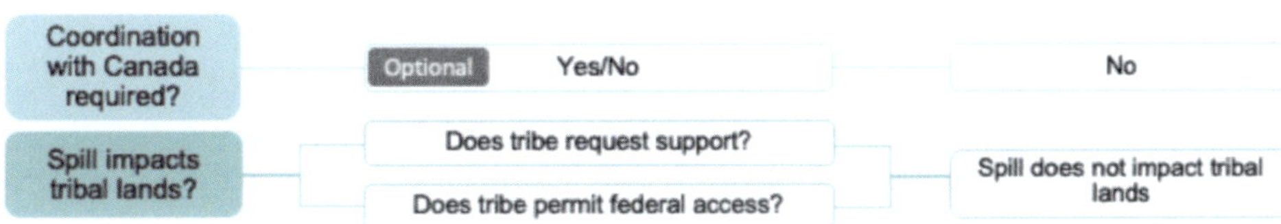

The next category is the type of shoreline habitat that will be at risk in the scenario. Shoreline habitat types dictate the response approach and equipment needed; for example, cleaning oil off a solid man-made structure requires a different approach and equipment than cleaning up a sandy beach or vegetation. Detailed shoreline habitat maps using Environmental Sensitivity Index (ESI) categories like the one shown in Figure B.6 are available in ERMA.

Figure B.6. Environmental Sensitivity Index Shoreline Habitat Map

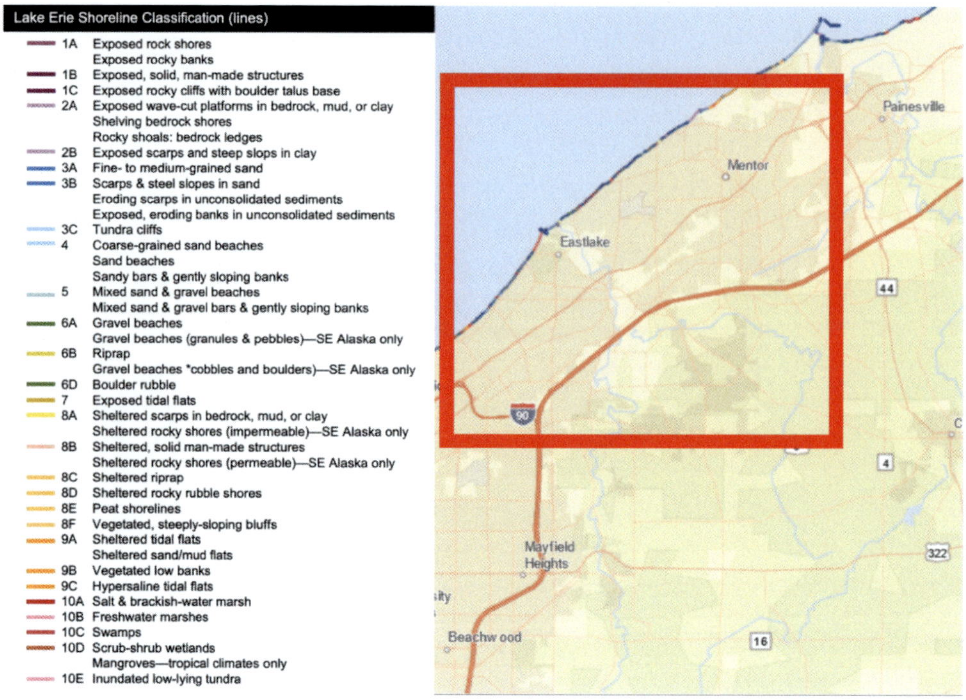

SOURCE: Adapted from ERMA, 2023.

For this location and scenario, there are multiple shoreline types. From the options shown in Figure B.7, we selected 1B exposed, solid man-made structures; 4 coarse grain sand beaches; 6A gravel beaches; 6B riprap revetments, groins, and jetties; and 10B fresh water marshes and 10C swamps as shoreline types that match our designated location.

Figure B.7. Filling in the Framework Template—Shoreline Habitat

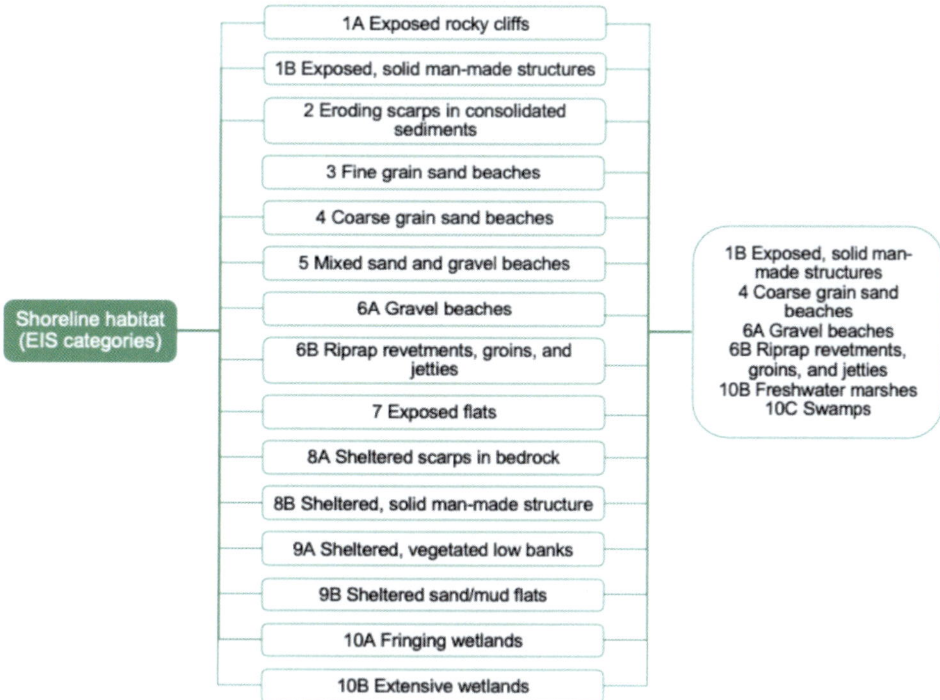

The next categories are the presence or absence of endangered species, historic or archaeological sites, or managed areas. For this scenario, the maps shown in Figures B.8, B.9, and B.10 reveal that none of these dimensions is present in the chosen location. The lack of these dimensions should be reflected in the template, as shown in Figure B.11.

Figure B.8. Subset of "Endangered Species Habitat Map"

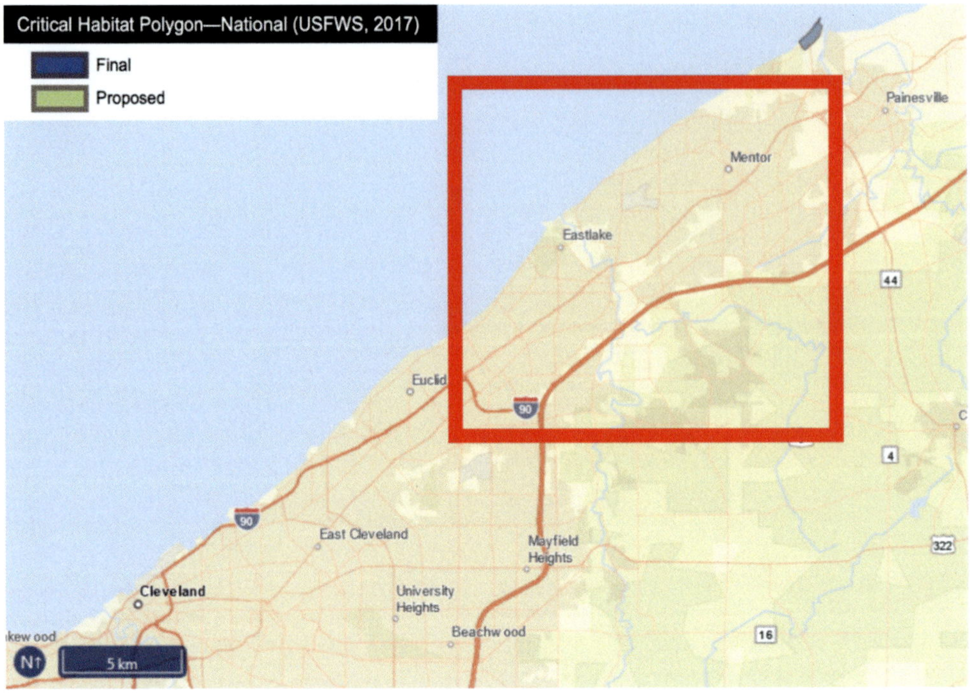

SOURCE: Adapted from ERMA, 2023.

Figure B.9. Subset of "National Register of Historic Places Map"

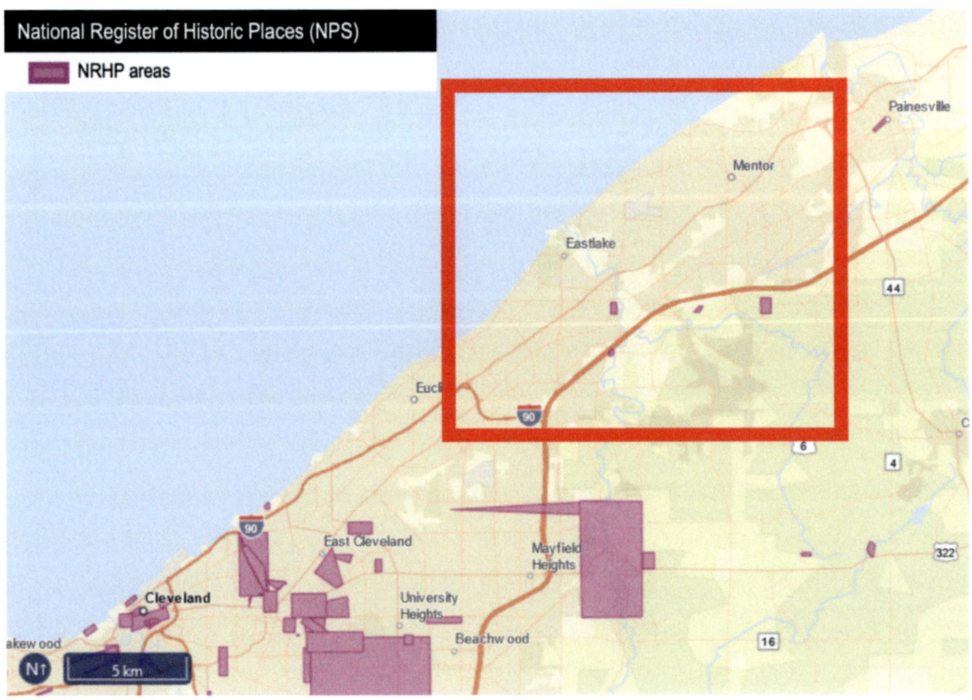

SOURCE: ERMA, 2023.

Figure B.10. Subset of "Marine Protected Areas and Wildlife Reserves Map"

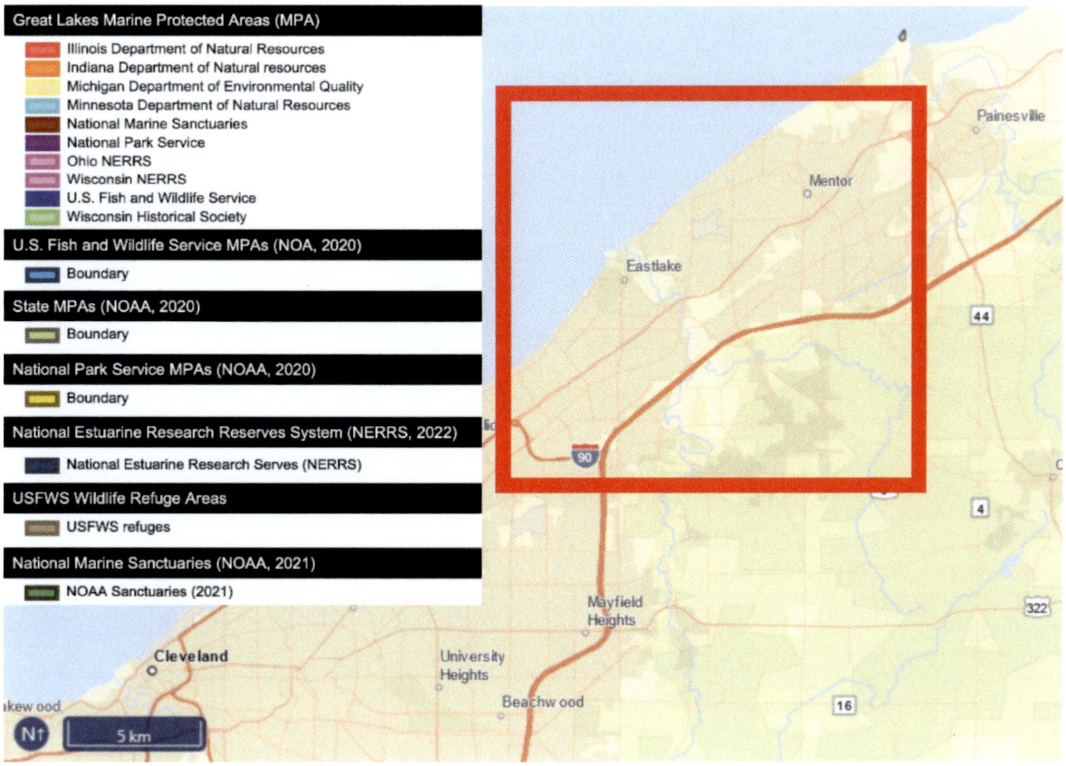

SOURCE: Adapted from ERMA, 2023.

Figure B.11. Filling in the Framework Template—Sensitive Areas

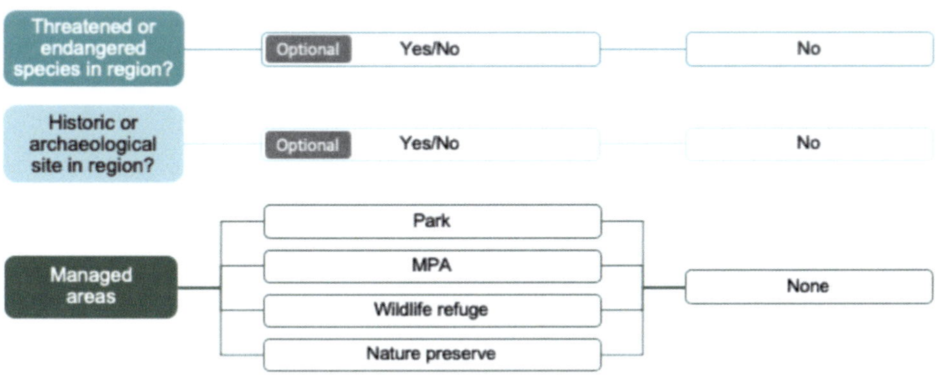

Step 4. Use the Framework to Choose Nongeographically Linked Factors, Considering Gaps and Less-Exercised Dimensions and Other Priorities

For this step, we choose factors for the scenario that are not linked to a specific geography. However, some factors are still related to other factors that *are* geographically linked. For example, the petroleum product type will relate to the source of the spill; not all spill sources carry all petroleum products. Therefore, as mentioned earlier, it is critical to elicit local expert opinions to ensure that the scenario is believable and realistic.

For this scenario, we are going to consider a medium discharge (1,000–10,000 gallons) of petroleum Group III (specific gravity 0.85–0.94), with the potential for additional discharge and no known responsible party because it is an orphan well of unknown size, as shown in Figure B.12. Note that, because this is an orphan well, we have no way of grounding this assumption with better data. This is a reasonable set of assumptions because it is likely that the USCG will not know what the oil type or potential size of the well spill could be unless the USCG already has contact with a local stakeholder who is expert in orphan oil wells.

Figure B.12. Filling in the Framework Template—Oil Outflow and Responsible Party

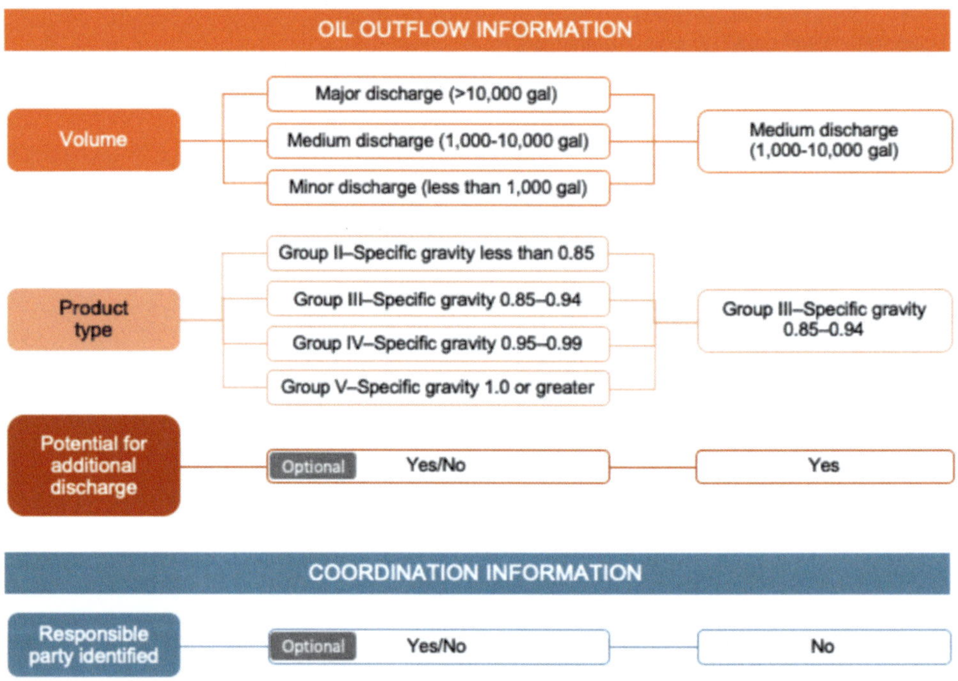

The Role of Weather in Scenarios

The next set of dimensions are all weather-related. If these scenarios are to be exercised in person, weather conditions should be chosen to reflect the reality of the weather on the date and time of the exercise. Because of this, it might be challenging (and unsafe) to practice oil-spill response in more extreme weather conditions, which might result in a persistent lack of response experience in extreme weather. This does not mean that users should avoid considering challenging weather conditions in scenario development; instead, they should consider challenging weather conditions as a factor that needs thorough discussion and pre-planning during scenario development and execution. For example, a scenario exercised in cold, clear weather could include ensuring that the appropriate equipment needed in snowy or icy conditions for safety, warmth, and response are available and in good condition, even if the weather during the exercise does not include snow or ice. If the scenarios are intended to be used for a tabletop exercise, the weather conditions can be more extreme. Discussions of planning and requirements should be documented as preparation for real-life challenging conditions.

Weather for Example Scenario

For this example, we consider conditions that would be challenging to re-create for an in-person exercise but that are useful considerations for response approaches. As shown in Figure B.13, we choose a rating between 2–4 on the Beaufort scale, meaning there is some wind and water movement but operating in this weather is not dangerous. Consequently, we do not include extreme or dangerous weather conditions in this scenario. However, we do include brash ice conditions: This scenario has 10–30 percent ice cover of less than two inches deep, meaning there are thin sheets and chunks of ice in the spill region. Water levels are average in this scenario.

Figure B.13. Filling in the Framework Template—Weather

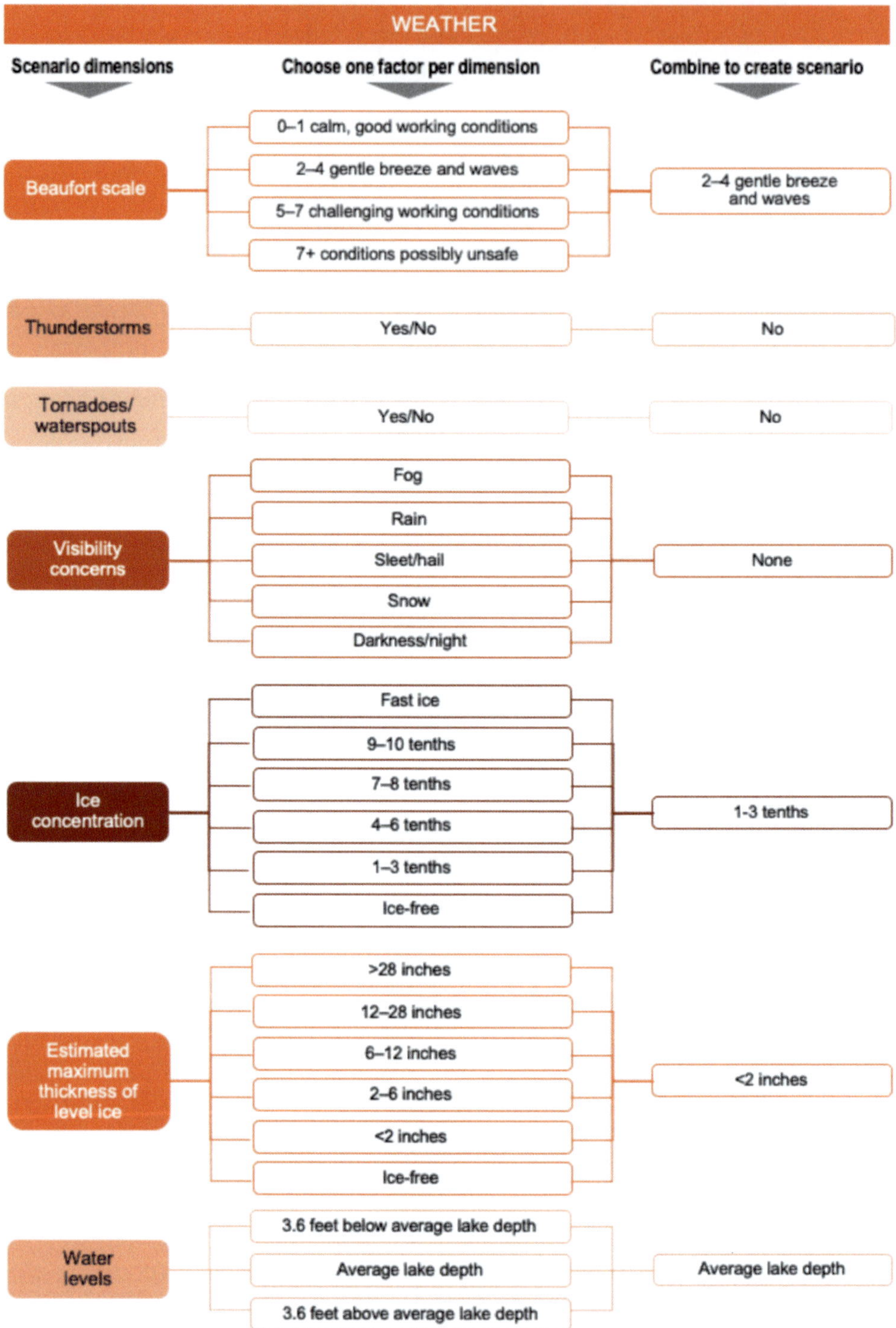

For the final two sections, Response Health and Safety Concerns and Additional Limitations, we can use the mapping of past exercises and regional priorities to develop the scenario. For this example, as shown in Figure B.14, we include two factors that limit workers' physical ability to respond: cold and poor visibility because of darkness. We do not include risk of injuries, fire, or terrorism/criminal factors. We are assuming that roadways are clear and accessible, but we have personnel shortages because of a holiday—for this scenario, the spill occurs on New Year's Day. There are no other major oil-spill or surge events that would draw additional USCG resources.

Figure B.14. Filling in the Framework Template—Health/Safety and Additional Limitations

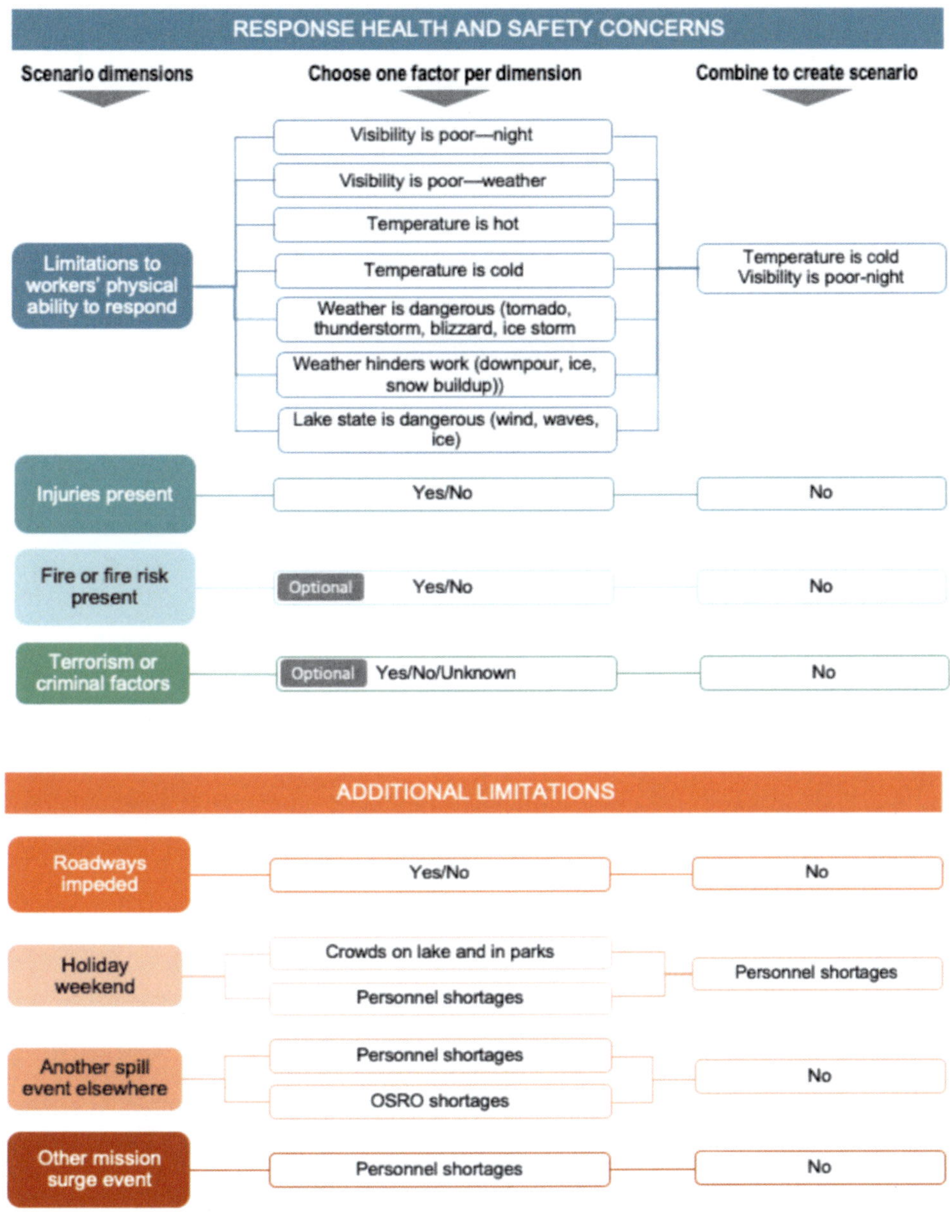

Step 5. Review Chosen Factors for Logic and Alignment with Scenario Priorities

The next step is to review all of the chosen factors to ensure they are logical and are in line with the goals of the scenarios or exercise. For this scenario, the factors that have been chosen are displayed in Figure B.15.

Figure B.15. Assessing the Scenario Template

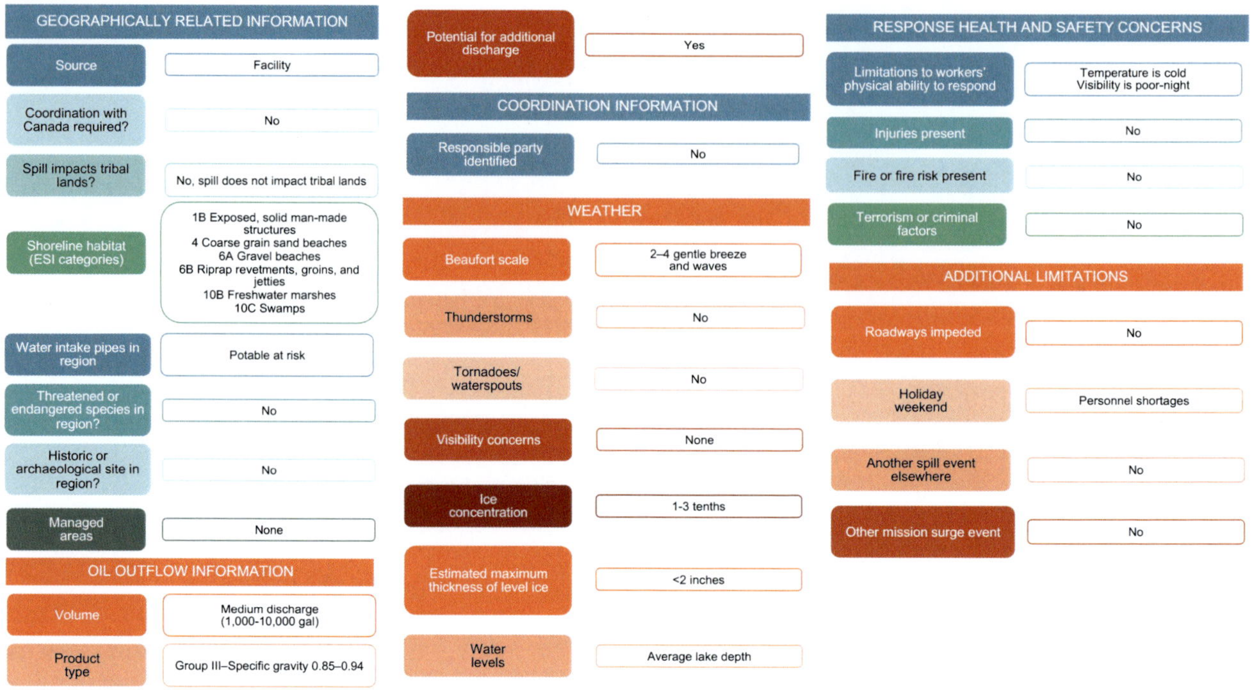

Considering the factors chosen for this scenario, we can see that the logic for this scenario works. There is a known risk of orphan wells in the chosen region; cold weather with a low Beaufort scale score and brash ice also makes sense. Setting the scenario on New Year's Day also means that shortened daylight hours, cold weather, and personnel shortages all make logical sense in the scenario.

Step 6. Develop a Narrative Around the Chosen Factors

Th narrative should be simple for the ease of both future replication and mapping into the suite of recently developed scenarios. These categories should drive the scenario narrative

- oil outflow information
- coordination information
- weather
- additional risks
- coastal environment
- response health and safety concerns
- additional limitations.

Example Scenario Narrative

On New Year's Day, coastal ice just east of Cleveland began to break up during a stretch of above-freezing weather. An oily sheen was reported, and on investigation, the source of the sheen was an orphaned oil well 20 yards from the shoreline. A recent spell of deep freezes followed by fast thaws might have weakened the old, degraded well infrastructure. The leak was not immediately detected because it was under ice for an unknown amount of time. The volume of the spill is suspected to be more than 1,000 gallons but less than 10,000 gallons. The potential for additional spill volume and the party responsible for the orphan well are unknown. The oil has been identified as Group 3.

The lake currently has minimal ice cover, but it is brash ice, so the ice chunks will make recovery challenging. The weather has been cold and clear, and on-water conditions are calm and good for response operations. The cold means that responders must be monitored for exposure.

The affected region does not have endangered species, managed lands, or historical sites, and no coordination is required with Canada or tribes. The oil must be cleaned from a variety of shorelines, which include manmade structures, coarse grain and gravel beaches, riprap, fresh water marshes, and swamps.

A major concern is that the spill appears to have affected a drinking water intake pipe, and it is unknown how long the pipe might have been exposed. Because of the holiday, there are personnel shortages of USCG and local water utility personnel. The holiday is also making it harder to reach offices that are not staffed 24/7. The only local stakeholder contact information available to USCG responders are the contacts listed in the ACP and GRS, and those will only be answered if they have been designated as 24/7 staffed on-call lines.

Step 7. Map Scenario Dimensions Onto the Suite of Recently Developed Scenarios

When the scenario has been developed and the exercise completed, the details of the scenarios can be added as a new column to Table B.1. This running list of exercises inventories the factors included in recent scenarios to prepare for developing the next scenario.

Step 8. Repeat

This process should be repeated and documented as needed.

Table B.1. Example Suite of After Action Reports Mapped Across 27 Dimensions

			After Action Report											
Category	**Dimension**	**Levers**	**Argo**	**LAK 21**	**LAK 22**	**JRT**	**HS**	**SL**	**USCG**	**SSM**	**SE**	**NOSMAC**	**TGV**	**#**
Oil outflow information	Source	Vessel-tanker	Yes	No	No	No	No	No	No	No	No	No	No	7
		Vessel-non-tanker	No	Yes	Yes	Yes	Yes	Yes	No	No	No	Yes	Yes	1
		Rail	No	No	No	No	No	No	No	Yes	No	No	No	0
		Truck	No	No	No	No	No	No	No	No	No	No	No	1
		Pipeline	No	No	No	No	No	No	No	No	Yes	No	No	0
		Facility	No	No	No	No	No	No	No	No	No	No	No	5
	Volume	Major discharge (>10,000 gal)	Yes	Yes	Yes	No	No	No	No	No	Yes	Yes	No	1
		Medium discharge (1,000–10,000 gal)	No	No	No	No	No	No	No	Yes	No	No	No	1
		Minor discharge (less than 1,000 gal)	No	No	No	No	No	No	Yes	No	No	No	No	2
	Product type	Group II: Specific gravity less than .85	No	Yes	No	No	No	No	Yes	No	No	No	No	3
		Group III: Specific gravity between .85 and less than .95	Yes	No	No	No	No	No	No	Yes	Yes	No	No	2
		Group IV:Specific gravity between .95 to and including 1.0	No	No	Yes	No	No	No	No	No	No	Yes	No	0
		Group V: Specific gravity greater than 1.0	No	No	No	No	No	No	No	No	No	No	No	7

			After Action Report											
Category	Dimension	Levers	Argo	LAK 21	LAK 22	JRT	HS	SL	USCG	SSM	SE	NOSMAC	TGV	#
	Potential for additional discharge	Yes	Yes	No	No	Yes	Yes	Yes	No	Yes	No	Yes	Yes	9
Coordination information	Responsible party identified	Yes	No	Yes	Yes	Yes	Yes	Yes	No	Yes	Yes	Yes	Yes	8
	Coordination with Canada required?	Yes	Yes	Yes	Yes	Yes	Yes	Yes	No	No	No	Yes	Yes	2
	Spill impacts Tribal lands	Was there coordination with Tribal entities? (U.S. only)	No	No	No	No	No	No	No	Yes	Yes	No	No	1
Weather	Beaufort scale (wind and waves)	0–1 calm, good working conditions	No	No	No	No	No	No	No	Yes	No	No	No	1
		2–4 gentle breeze and waves	No	No	No	No	No	No	Yes	No	No	No	No	0
		5–7 challenging working conditions	No	No	No	No	No	No	No	No	No	No	No	0
		7+ conditions possibly unsafe	No	No	No	No	No	No	No	No	No	No	No	0
	Thunderstor ms	Yes	No	No	No	No	No	No	No	No	No	No	No	0
	Tornadoes/w aterspouts	Yes	No	No	No	No	No	No	No	No	No	No	No	0
	Visibility concerns	Fog	No	No	No	No	No	No	No	No	No	No	No	1
		Rain	No	No	No	No	No	No	No	Yes	No	No	No	0
		Sleet/hail	No	No	No	No	No	No	No	No	No	No	No	1
		Snow	No	No	No	No	No	No	Yes	No	No	No	No	0
		Darkness/night	No	No	No	No	No	No	No	No	No	No	No	1

			After Action Report											
Category	Dimension	Levers	Argo	LAK 21	LAK 22	JRT	HS	SL	USCG	SSM	SE	NOSMAC	TGV	#
	Ice presence?	Yes	No	No	No	No	No	No	Yes	No	No	No	No	0
	Ice concentration	Fast Ice	No	No	No	No	No	No	No	No	No	No	No	0
		9–10 Tenths	No	No	No	No	No	No	No	No	No	No	No	0
		7–8 Tenths	No	No	No	No	No	No	No	No	No	No	No	0
		4–6 Tenths	No	No	No	No	No	No	No	No	No	No	No	0
		1–3 Tenths	No	No	No	No	No	No	No	No	No	No	No	3
		Ice Free	No	No	No	No	No	No	No	Yes	Yes	No	Yes	0
	Estimated Maximum Thickness of Level Ice	>28 inches	No	No	No	No	No	No	No	No	No	No	No	0
		12–28 inches	No	No	No	No	No	No	No	No	No	No	No	1
		6–12 inches	No	No	No	No	No	No	Yes	No	No	No	No	1
		2–6 inches	No	No	No	No	No	No	Yes	No	No	No	No	0
		<2 inches	No	No	No	No	No	No	No	No	No	No	No	3
		Ice Free	No	No	No	No	No	No	No	Yes	Yes	No	Yes	0
	Water levels	3.6' below average lake depth	No	No	No	No	No	No	No	No	No	No	No	0
		average lake depth	No	No	No	No	No	No	No	No	No	No	No	0
		3.6' above average lake depth	No	No	No	No	No	No	No	No	No	No	No	0
Additional risks	Water intake pipes in region	Potable impacted	No	No	No	No	No	No	No	No	No	No	No	0
		Potable at risk	No	No	No	No	No	No	No	No	No	No	No	0
		Non-potable impacted	No	No	No	No	No	No	No	No	No	No	No	0

			After Action Report											
Category	Dimension	Levers	Argo	LAK 21	LAK 22	JRT	HS	SL	USCG	SSM	SE	NOSMAC	TGV	#
		Non-potable at risk	No	No	No	No	No	No	No	No	No	No	No	0
	Prescence of/habitat for federally threatened or endangered species	Yes	No	No	No	No	No	No	No	No	No	No	No	0
	Historic or archaeological site in region	Yes	No	No	No	No	No	No	No	No	No	No	No	0
Coastal environment	Shoreline habitat rankings (EIS categories)	1A Exposed Rocky Cliffs	No	No	No	No	No	No	No	No	No	No	No	0
		1B Exposed, solid man-made structures	No	No	No	No	No	No	No	No	No	No	No	0
		2 Shelving bedrock shores	No	No	No	No	No	No	No	No	No	No	No	0
		3 Eroding scarps in consolidated sediments	No	No	No	No	No	No	No	No	No	No	No	0
		4 Sand beaches	No	No	No	No	No	No	No	No	No	No	No	0
		5 Mixed sand and gravel beaches	No	No	No	No	No	No	No	No	No	No	No	0
		6A Gravel beaches	No	No	No	No	No	No	No	No	No	No	No	0
		6B Riprap revetments, groins and jetties	No	No	No	No	No	No	No	No	No	No	No	0
		7 Exposed flats	No	No	No	No	No	No	No	No	No	No	No	0
		8A Sheltered scarps in bedrock	No	No	No	No	No	No	No	No	No	No	No	0

			After Action Report											
Category	Dimension	Levers	Argo	LAK 21	LAK 22	JRT	HS	SL	USCG	SSM	SE	NOSMAC	TGV	#
		8B Sheltered, solid man-made structured	No	No	No	No	No	No	No	No	No	No	No	0
		9A Sheltered, vegetated low banks	No	No	No	No	No	No	No	No	No	No	No	0
		9B Sheltered sand/mud flats	No	No	No	No	No	No	No	No	No	No	No	0
		10A Fringing wetlands	No	No	No	No	No	No	No	No	No	No	No	2
		10B Extensive wetlands	No	Yes	Yes	No	No	No	No	No	No	No	No	0
	Managed area	Park	No	No	No	No	No	No	No	No	No	No	No	0
		MPA	No	No	No	No	No	No	No	No	No	No	No	2
		Wildlife refuge	No	Yes	Yes	No	No	No	No	No	No	No	No	0
		Nature Preserve	No	No	No	No	No	No	No	No	No	No	No	0
Response health and safety concerns	Limitations to workers' physical ability to respond	Visibility is poor–nighttime	No	No	No	No	No	No	No	No	No	No	No	1
		Visibility is poor–weather	No	No	No	No	No	No	Yes	No	No	No	No	0
		Temperature is too hot	No	No	No	No	No	No	No	No	No	No	No	1
		Temperature is too cold	No	No	No	No	No	No	Yes	No	No	No	No	1
		Weather is dangerous (tornado, thunderstorm, blizzard, ice storm)	No	No	No	No	No	No	Yes	No	No	No	No	2
		Weather hinders work (downpour, ice, snow buildup)	Yes	No	No	No	No	No	Yes	No	No	No	No	1

			After Action Report											
Category	Dimension	Levers	Argo	LAK 21	LAK 22	JRT	HS	SL	USCG	SSM	SE	NOSMAC	TGV	#
		Lake state is dangerous (wind, waves, ice)	No	No	No	No	No	No	Yes	No	No	No	No	0
	Injuries present	Yes	No	No	No	No	No	No	No	No	No	No	No	1
	Fire or fire risk present	Yes	No	No	No	No	No	No	Yes	No	No	No	No	0
	Terrorism or criminal factors	Yes	No	No	No	No	No	No	No	No	No	No	No	0
		Unknown	No	No	No	No	No	No	No	No	No	No	No	0
Additional limitations	Roadways impeded	Yes	No	No	No	No	No	No	No	No	No	No	No	0
	Holiday weekend	Crowds on lake and in parks	No	No	No	No	No	No	No	No	No	No	No	0
		Personnel shortages	No	No	No	No	No	No	No	No	No	No	No	0
	Other spill event already underway elsewhere	Personnel shortages	No	No	No	No	No	No	No	No	No	No	No	0
		OSRO shortages	No	No	No	No	No	No	No	No	No	No	No	0
	Other mission surge event	Personnel shortages	No	No	No	No	No	No	No	No	No	No	No	0

SOURCE: Features information from Canadian Coast Guard Environmental Response, *Great Lakes Joint Response Team Harvest Spirit Grounding After Action Report: 02 Dec–04 Dec 2020*, provided to authors, June 1, 2021a; Canadian Coast Guard Environmental Response, 2021b; Canadian Coast Guard Environmental Response, *CANUSLAK 2022 Exercise After Action Report: 17 May 2022*, provided to authors, July 7, 2022a; Canadian Coast Guard Environmental Response, *Great Lakes Joint Response Team Barge PML 9000 Grounding After Action Report 04 May–07 May 2022*, provided to authors, August 25, 2022b; Canadian Coast Guard Environmental Response, 2022c; Canadian Coast Guard Environmental Response, *Great Lakes Joint Response Team M/V SAM LAUD Grounding After Action Report: 08 Oct–09 Oct 2022*, provided to authors, January 18, 2023; Pace, 2023; USCG, undated a; U.S. Coast Guard, *M/V Harvest Spirit Marine Transportation Disruption After Action Report: 02 Dec 2020–04 Dec 2020*, provided to authors, undated b; USCG, undated c; USCG, 2013; USCG, 2018b; USCG, 2019a.

NOTE: PREP = Preparedness for Response Exercise Program.
Argo = Argo After-Action Report
LAK 21 = CANUSLAK 2021

LAK 22 = CANUSLAK 2022
JRT = Great Lakes Joint Response Team 2022
HS = Harvest Spirit Grounding AAR and MV Harvest Spirit
SL = M/V Sam Laud Grounding 2022
USCG = USCG GL Oil Spill Response Exercise
SSM = Manistique ICS 201 and Sector SSM PREP FSE 2018
SE = Straits exercise
NOSMAC = NOSMAC PREP Full-Scale Exercise
TGV = Tecumseh Ground Vessel 2022
# = # of times dimension has been exercised recently.

Figure B.16. Scenario Development Template with Full Dimensions and Factors

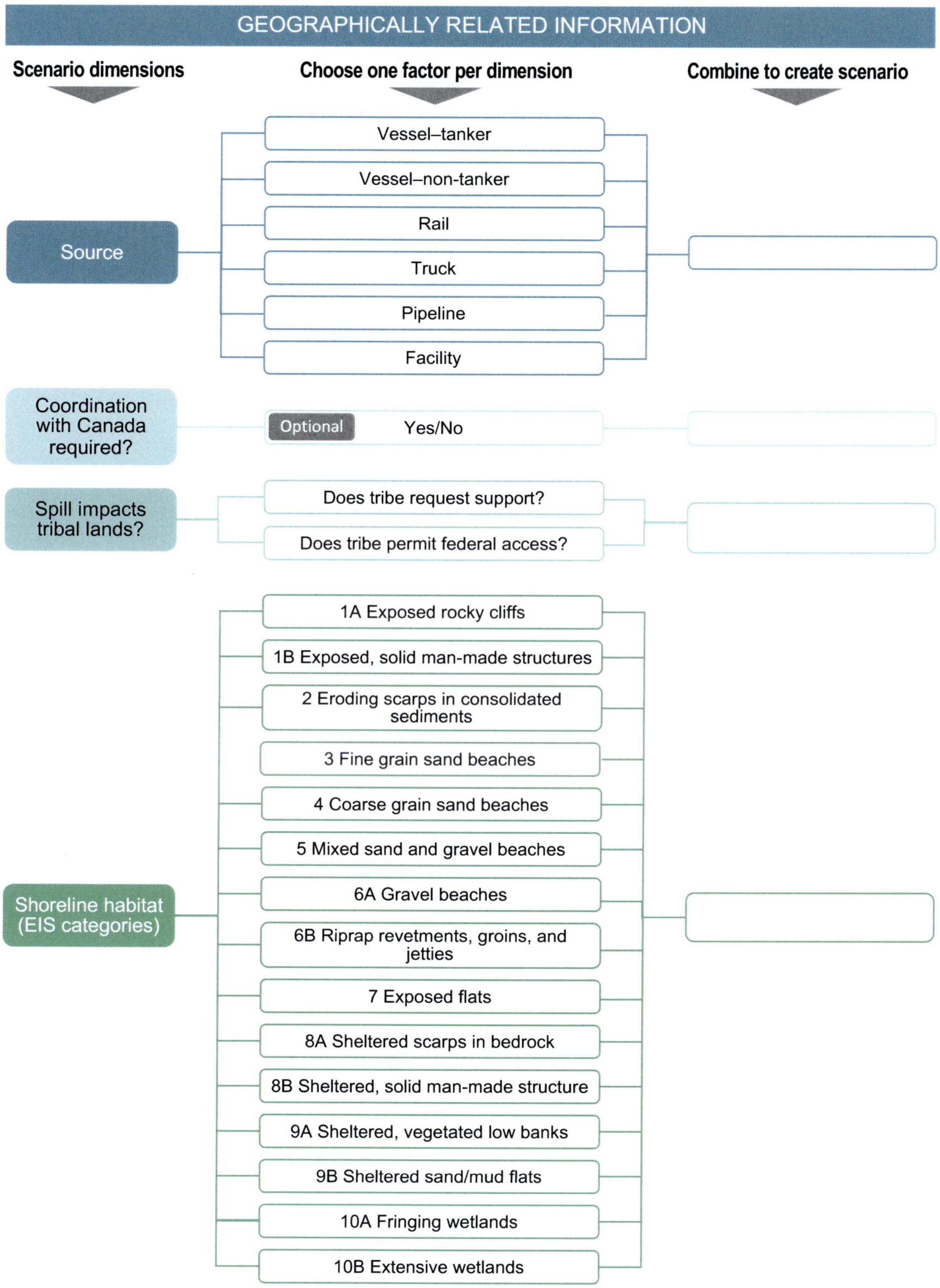

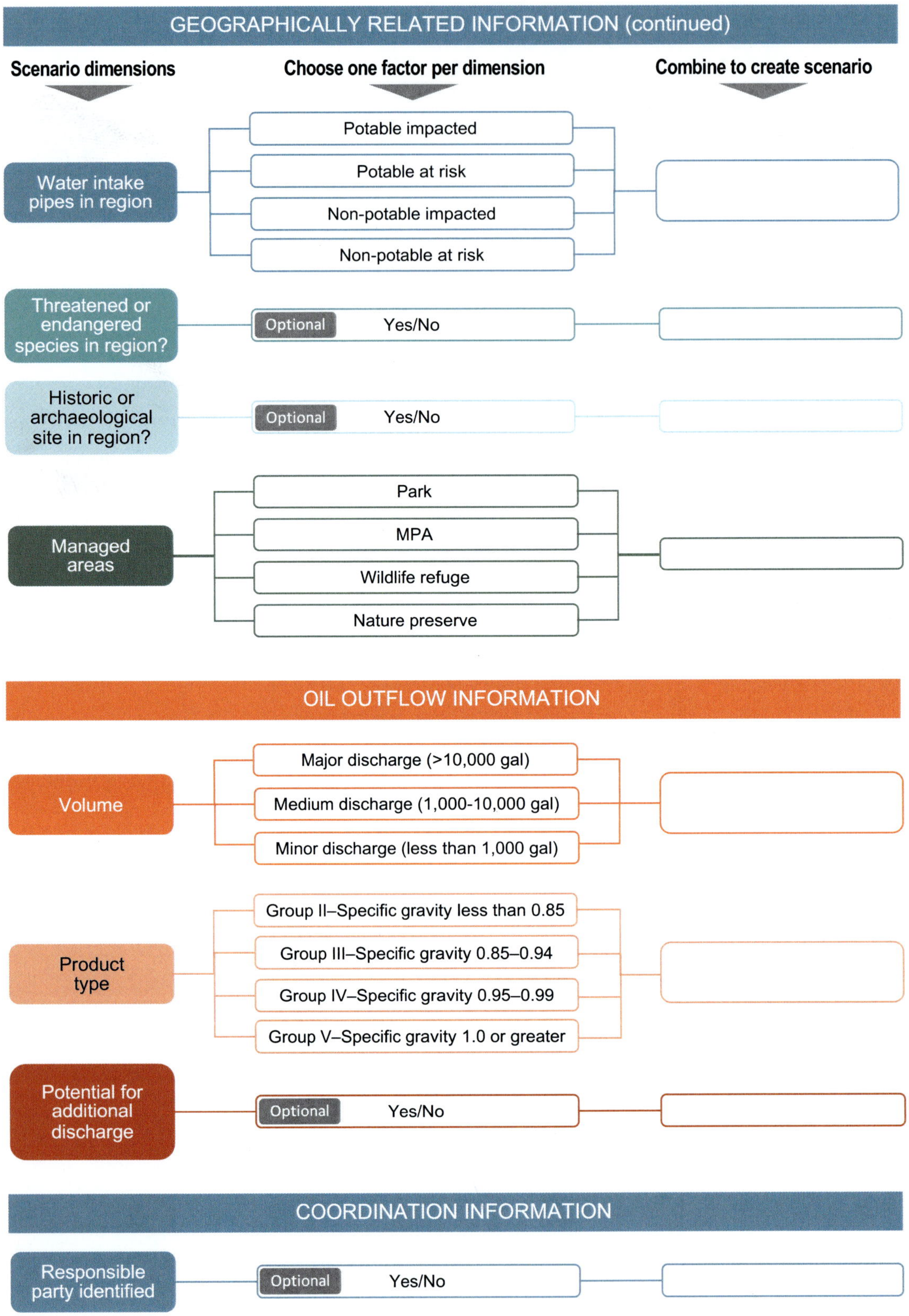
GEOGRAPHICALLY RELATED INFORMATION (continued)
Scenario dimensions
Choose one factor per dimension
Combine to create scenario
Water intake pipes in region
Potable impacted
Potable at risk
Non-potable impacted
Non-potable at risk
Threatened or endangered species in region?
Optional
Yes/No
Historic or archaeological site in region?
Optional
Yes/No
Managed areas
Park
MPA
Wildlife refuge
Nature preserve
OIL OUTFLOW INFORMATION
Volume
Major discharge (>10,000 gal)
Medium discharge (1,000-10,000 gal)
Minor discharge (less than 1,000 gal)
Product type
Group II–Specific gravity less than 0.85
Group III–Specific gravity 0.85–0.94
Group IV–Specific gravity 0.95–0.99
Group V–Specific gravity 1.0 or greater
Potential for additional discharge
Optional
Yes/No
COORDINATION INFORMATION
Responsible party identified
Optional
Yes/No

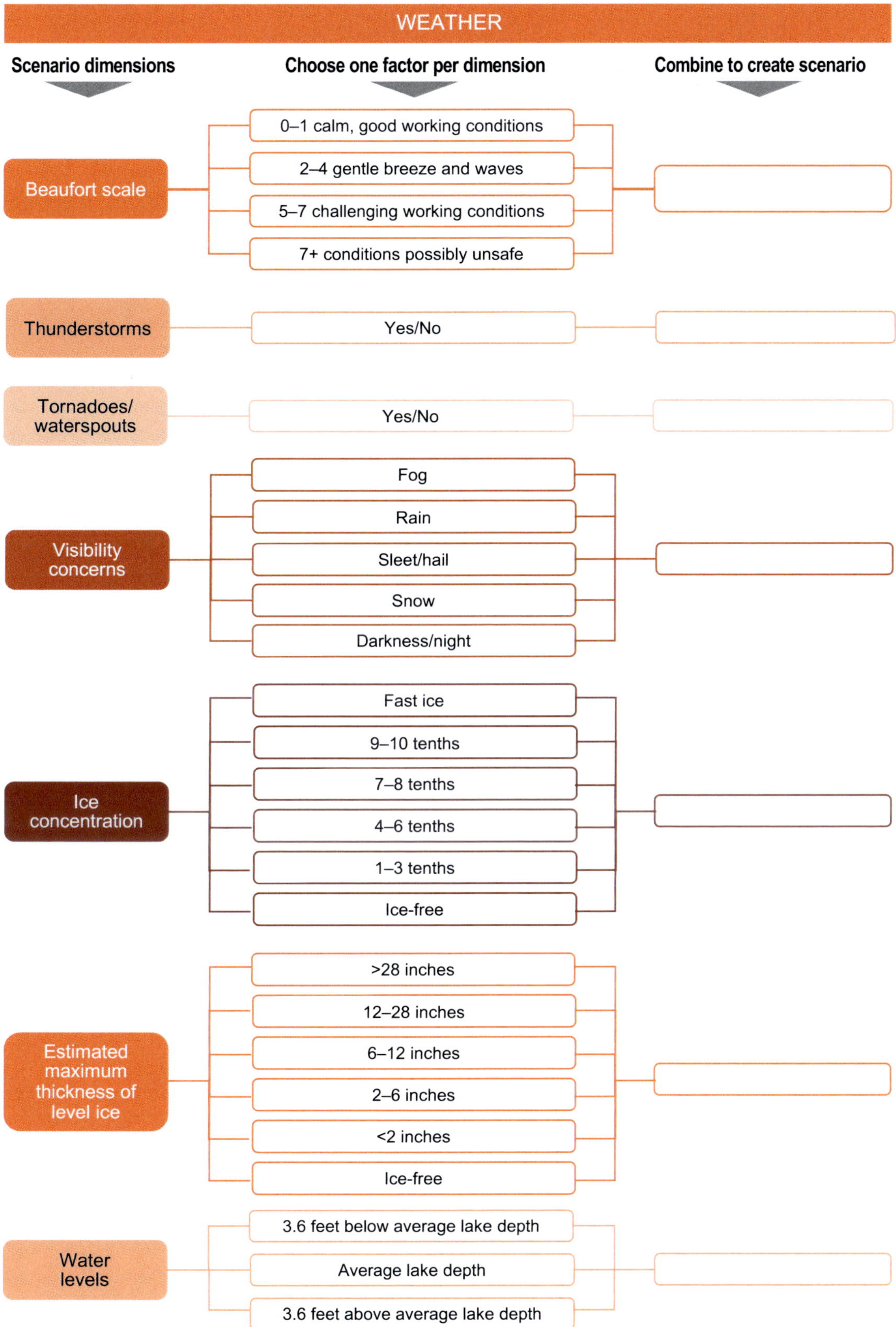
WEATHER
Scenario dimensions
Choose one factor per dimension
Combine to create scenario
Beaufort scale
0–1 calm, good working conditions
2–4 gentle breeze and waves
5–7 challenging working conditions
7+ conditions possibly unsafe
Thunderstorms
Yes/No
Tornadoes/ waterspouts
Yes/No
Visibility concerns
Fog
Rain
Sleet/hail
Snow
Darkness/night
Ice concentration
Fast ice
9–10 tenths
7–8 tenths
4–6 tenths
1–3 tenths
Ice-free
Estimated maximum thickness of level ice
>28 inches
12–28 inches
6–12 inches
2–6 inches
<2 inches
Ice-free
Water levels
3.6 feet below average lake depth
Average lake depth
3.6 feet above average lake depth

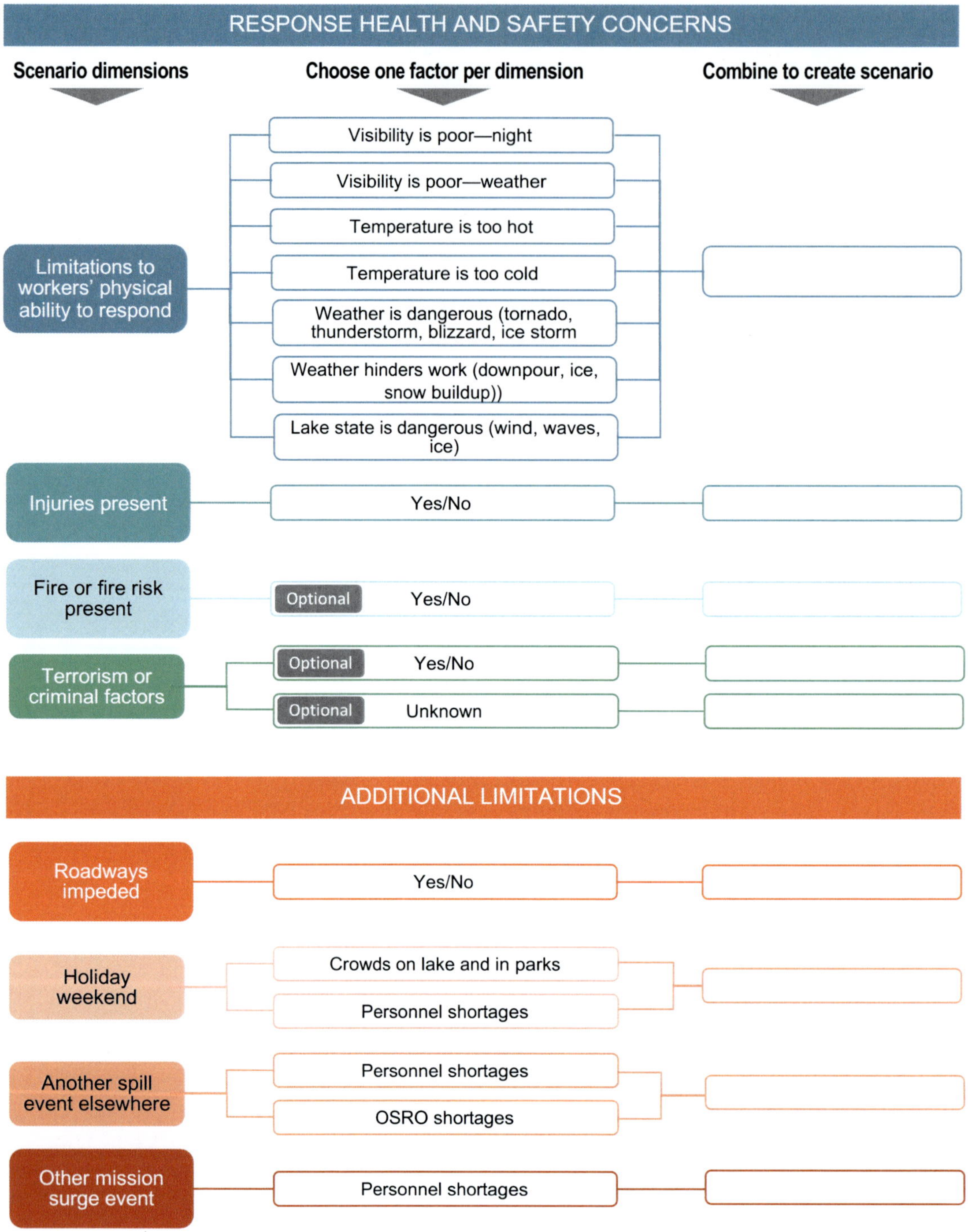
RESPONSE HEALTH AND SAFETY CONCERNS
Scenario dimensions
Choose one factor per dimension
Combine to create scenario
Limitations to workers' physical ability to respond
Visibility is poor—night
Visibility is poor—weather
Temperature is too hot
Temperature is too cold
Weather is dangerous (tornado, thunderstorm, blizzard, ice storm
Weather hinders work (downpour, ice, snow buildup))
Lake state is dangerous (wind, waves, ice)
Injuries present
Yes/No
Fire or fire risk present
Optional
Yes/No
Terrorism or criminal factors
Optional
Yes/No
Optional
Unknown
ADDITIONAL LIMITATIONS
Roadways impeded
Yes/No
Holiday weekend
Crowds on lake and in parks
Personnel shortages
Another spill event elsewhere
Personnel shortages
OSRO shortages
Other mission surge event
Personnel shortages

SOURCES: ERMA, 2023; Enbridge, 2023; GRS, "Algoma, WI: WLM17-ALG," USGS, undated a; GRS, "Baileys Harbor, WI: WLM24-BAL," USGS, undated b; GRS, "Burns Harbor, IN: SLM1-BUR," USGS, undated c; GRS, "Chicago, IL: SLM7-CHI," USGS, undated d; GRS, "East Chicago, IN: SLM4-ECH," USGS, undated e; GRS, "Ephraim Bay, WI: WLM32-EPH," USGS, undated f; GRS, "Escanaba/Little Bay DeNoc: NLM5-ESC," USGS, undated g; GRS, "Evanston, IL: SLM8-EVA," USGS, undated h; GRS, "Frankfort, MI: ESI# ELM1-FRA," USGS, undated i; GRS, "Grand Haven/Port Sheldon: ELM09-GRH," USGS, undated j; GRS, "Green Bay, WI: WLM42-GBS," USGS, undated k; GRS, "Holland and Saugatuk: ELM11-HOL," USGS, undated l; GRS, "Jacksonport, WI: WLM22-JAX," USGS, undated m; GRS, "Kenosha, WI: WLM1-KEN," USGS, undated n; GRS, "Little Sable Pt./Muskegon/White Lake: ELM07-MUS," USGS, undated o; GRS, "Little Sturgeon Bay, WI: WLM38-LSB," USGS, undated p; GRS, "Ludington, MI: ESI# - ELM3-LUD," USGS, undated q; GRS, "Manistee, MI: ESI# - ELM2-MAN," USGS, undated r; GRS, "Manitowoc, WI: WLM13-MAN," USGS, undated s; GRS, "Marinette, WI: WLM51-MAR," USGS, undated t; GRS, "Michigan City/New Buffalo: ELM23-MIC," USGS, undated u; GRS, "Milwaukee, WI: WLM5-MKE," USGS, undated v; GRS, "Port Washington, WI: WLM8-PWA," USGS, undated w; GRS, "Racine, WI: WLM2-RAC," USGS, undated x; GRS, "Sector Detroit GRS Div 3: Bay City," USGS, undated y; GRS, "Sector Detroit GRS Div 4: Saginaw Bay," USGS, undated z; GRS, "Sector Detroit Lake Huron Division 1," USGS, undated aa; GRS, "Sector Detroit Saginaw Bay Division 2," USGS, undated bb; GRS, "Sheboygan, WI: WLM10-SHE," USGS, undated cc; GRS, "Sister Bay, WI: WLM24-SIS," USGS, undated dd; GRS, "Sleeping Bear Dunes: NLM64-SBD," USGS, undated ee; GRS, "South Haven: ELM14-SHA," USGS, undated ff; GRS, "St Joseph: ELM17-STJ," USGS, undated gg; GRS, "Sturgeon, Bay, WI: WLM20-STB," USGS, undated hh; GRS, "Two Rivers, WI: WLM14-TWO," USGS, undated ii; GRS, "Washington Island, WI: WLM27-WAS," USGS, undated jj; GRS, "Waukegan, IL: WLM11-WAK," USGS, undated kk; Chelsey Lewis, "Huge Waves Hit Lake Michigan's Eastern Shore, Waves Set Record in October," *Milwaukee Journal Sentinel*, December 7, 2017; NOAA, "GLCFS Modeled NOWCAST Wave Heights vs Observed Wave Heights," webpage, 2017; NOAA, "Storm Events Database," webpage, undated; NOAA, Great Lakes Environmental Research Laboratory, "Great Lakes Water Level Observations," undated a; Jerome A. Popiel, "Sector Detroit ACP.pdf," email to K. Anania, C. Pace K., C. J. Graham, and A. Davenport, USCG, February 7, 2023; USCG, 2015; USCG, *Eastern Great Lakes Area Contingency Plan Geographic Response Plan for Lake Ontario*, 2017a; USCG, *Eastern Great Lakes Area Contingency Plan Geographic Response Plan for Northeast Ohio*, 2017b; USCG, *Eastern Great Lakes Area Contingency Plan, Sector Buffalo*, 2017e; USCG, *Eastern Great Lakes Area Contingency Plan: Geographic Response Plan for Northwestern Pennsylvania*, 2017c; USCG, *Eastern Great Lakes Area Contingency Plan: Geographic Response Plan for St Lawrence Seaway*, 2017d; USCG, *Northern Michigan Area Contingency Plan*, 2021b; USCG, *Northern Michigan Area Contingency Plan, Sector Sault Sainte Marie*, 2017g; USCG, *Northwest Ohio & Southeast Michigan Area Contingency Plan, Sector Detroit*, May 13, 2019b; USCG, *Coast Guard Sector Lake Michigan Area Contingency Plan*, May 14, 2019c; USCG, *Western Lake Superior Area Contingency Plan*, MSU Duluth, October 28, 2016; U.S. National Ice Center, "Great Lakes: Ice Charts," webpage, undated.

Reference Maps for Geographically Linked Dimensions

The maps in this section each show a single relevant feature of the Great Lakes region. These maps are used for the scenario-building process described in Figure B.16. If the location of a scenario is planned before the dimensions of the scenario are chosen, these maps show which exercise dimensions are realistic for that location. Alternatively, if the geographically linked dimension(s) of an exercise are chosen before the location, looking at maps showing the feature(s) of interest can help to choose a realistic location for the exercise. The maps in Figures B.17–B.23 correspond to the source, coordination with tribal lands,[191] shoreline habitats, water intake pipes, threatened or endangered species, historical or archaeological sites, and managed areas dimensions in Figure B.16.

There are two mapping applications that are helpful when building a scenario in the Great Lakes. For the source dimensions in Figure B.16, the EIA Atlas mapping application shows where oil is moved via rail and pipeline, and the locations of facilities that handle oil and could be sources of oil spills.[192] The facilities in the application are oil wells, crude oil and petroleum product pipelines, crude oil and petroleum product rail terminals, petroleum refineries, and the ports and waterways that petroleum products are shipped through. Oil tanker and other cargo vessel shipping tracks cannot be found on this application. Although it does not host a mapping application, Marine Cadastre, a joint BOEM and NOAA initiative, has geographical data showing the AIS tracks of cargo vessels in the Great Lakes.[193] The other geographically linked dimensions can be found on NOAA's Great Lakes environmental response management application.[194]

---

[191] Tribal lands are shown in the figures using dark green shading.

[192] U.S. Energy Information Administration, undated.

[193] NOAA and BOEM, undated.

[194] ERMA, accessed June 1, 2023.

Figure B.17. Fixed Oil Spill Risk Sources Within Five Miles of Coast

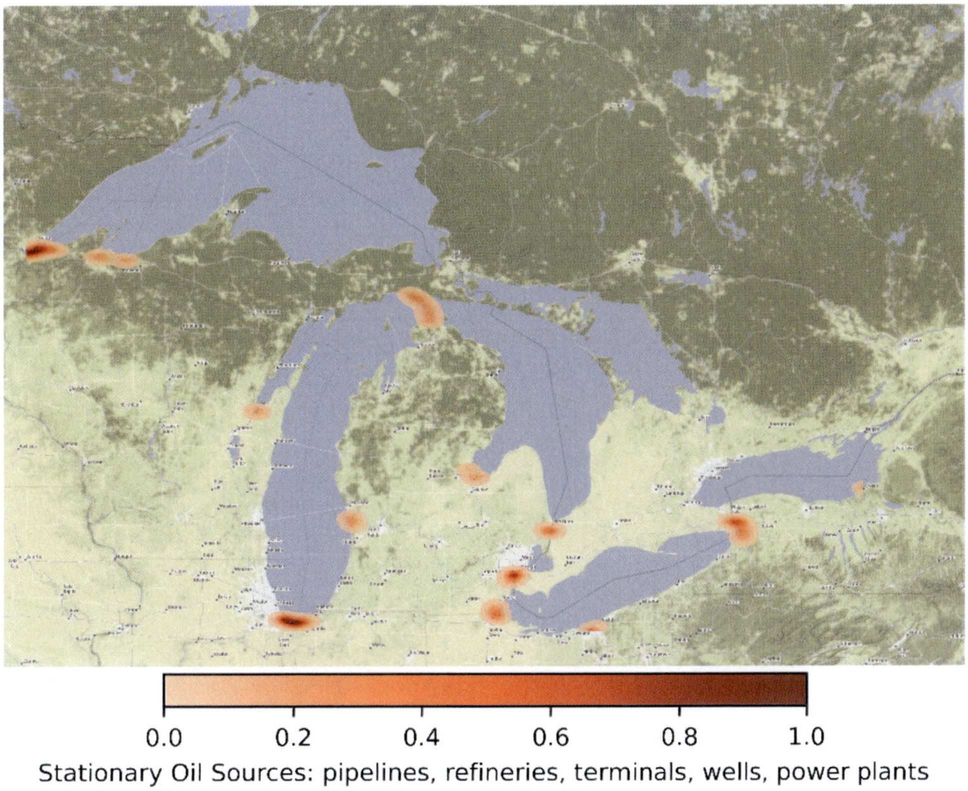

SOURCE: Features information from U.S. Energy Information Administration, undated.

Figure B.18. Moveable Sources of Oil Spill Risk Within Five Miles of Coast

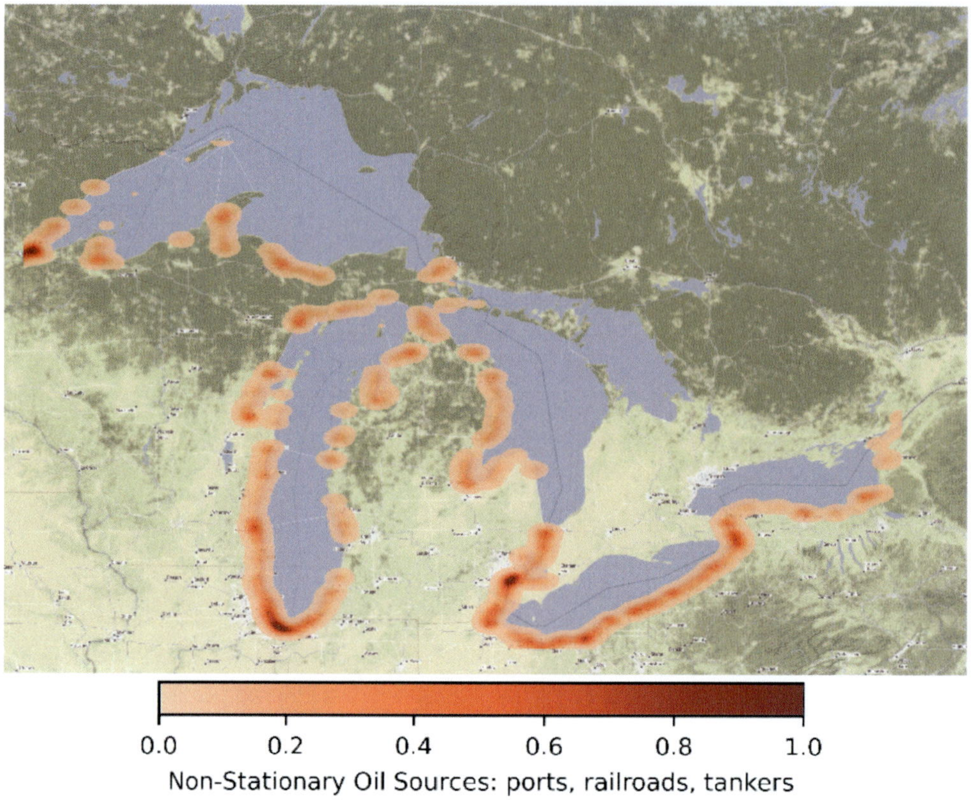

SOURCE: Features information from U.S. Energy Information Administration, undated; NOAA and BOEM, undated.

Figure B.19. Map of Tribal Lands Within Five Miles of Coast

SOURCE: Features information from ERMA, 2023.

Figure B.20. Shoreline Classifications from ERMA

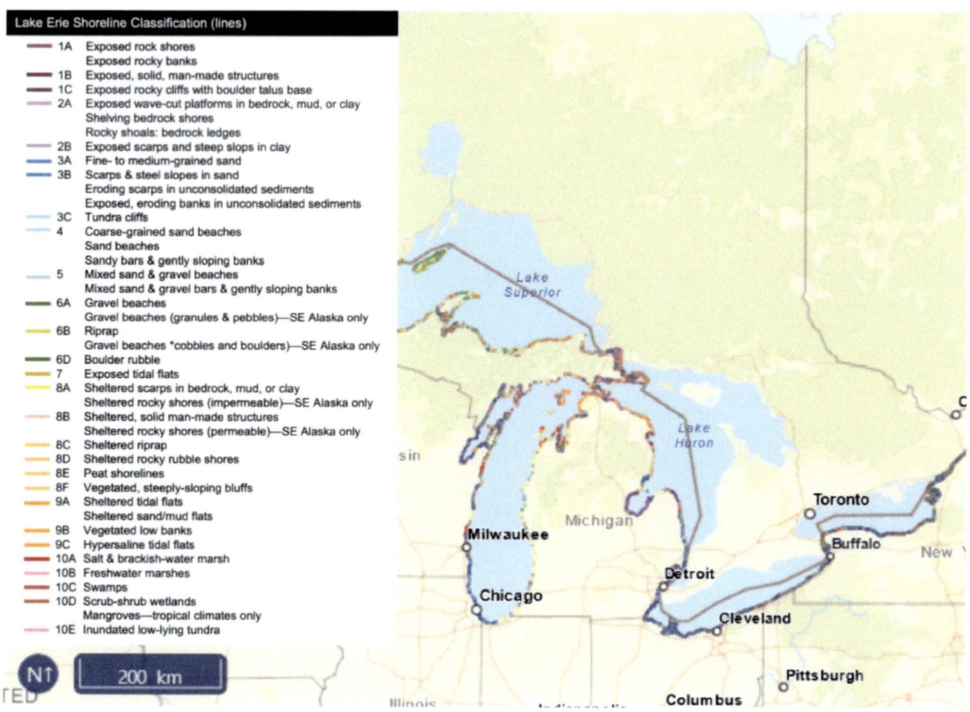

SOURCE: Reproduced from ERMA, 2023.

Figure B.21. Shoreline Classification from ERMA, Zoomed in

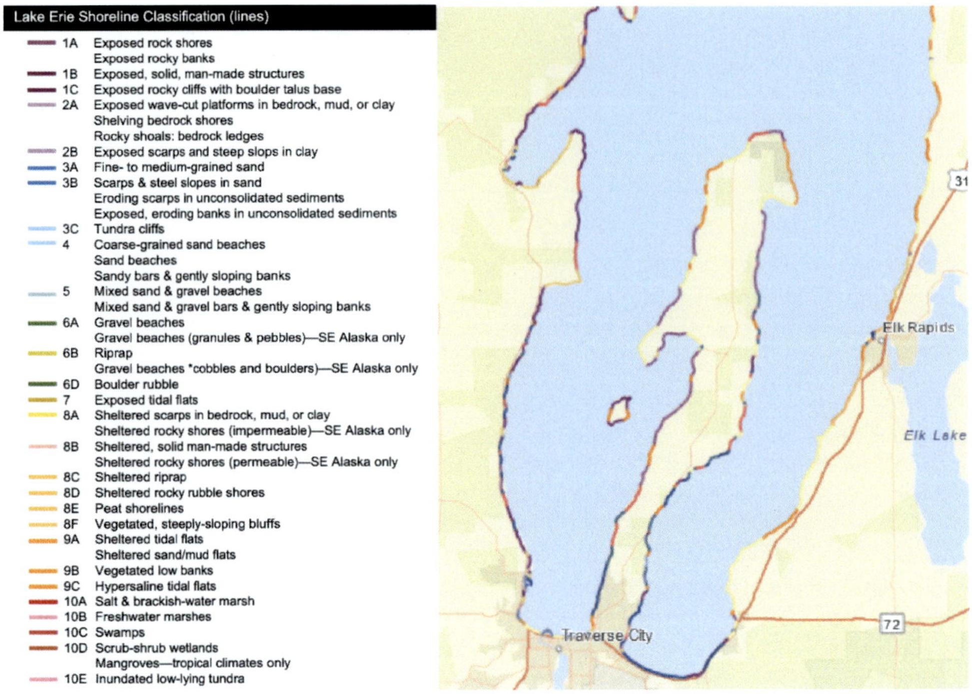

SOURCE: Reproduced from ERMA, 2023.

Figure B.22. Public Water System Drinking Water Intakes in Ohio

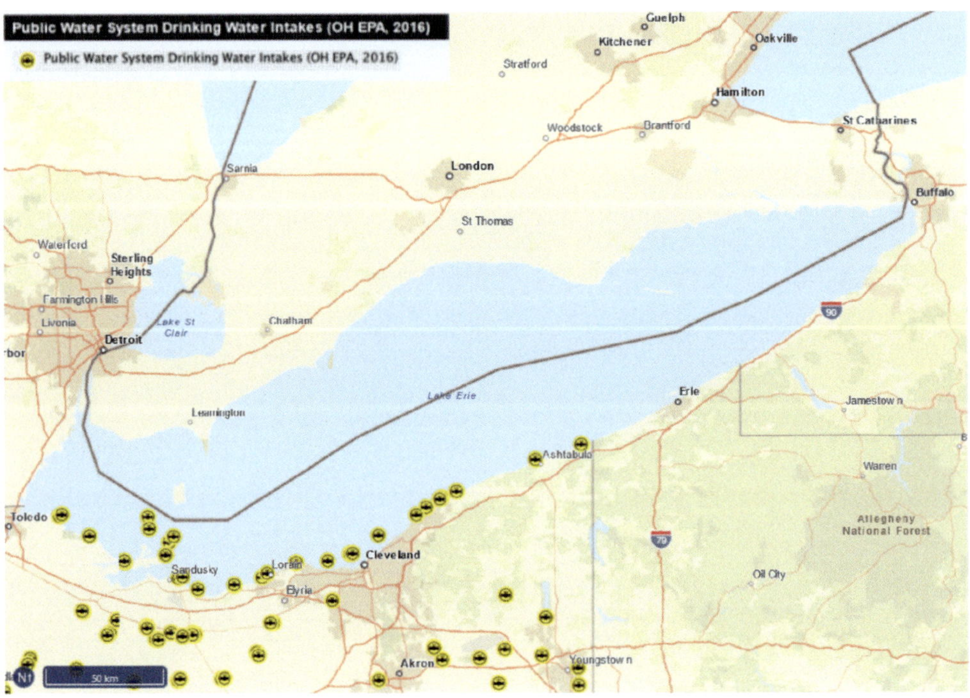

SOURCE: Reproduced from ERMA, 2023.

Figure B.23. Critical Habitats in the Great Lakes Region

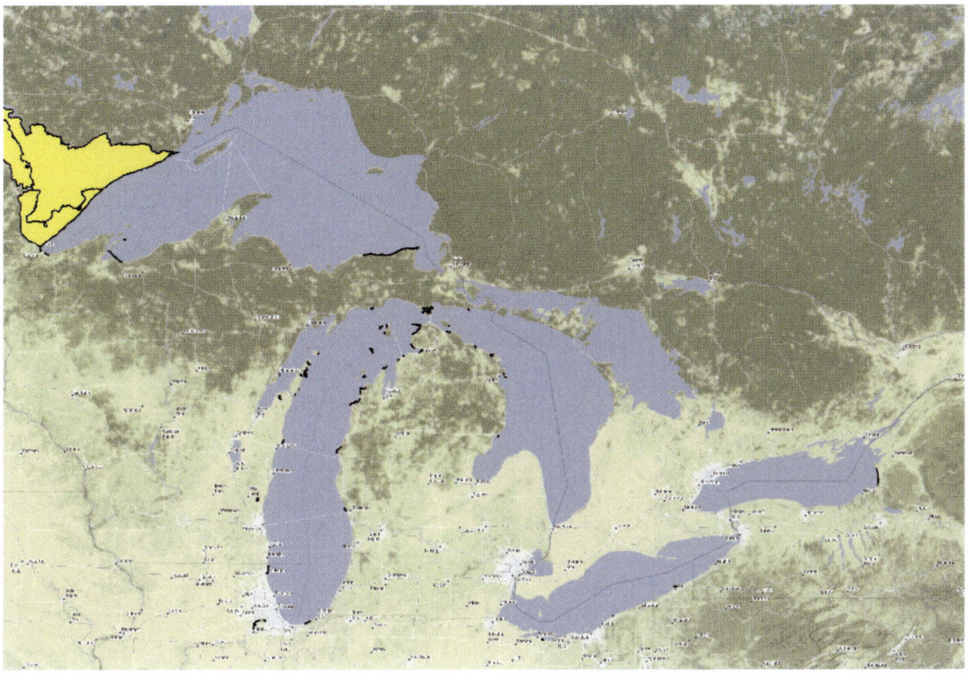

SOURCE: Features information from ERMA, 2023.

Sector-Specific Oil Spill Risk Dimensions

Each sector has its own ACP, and with the exception of Sector Buffalo, each ACP has a description of worst case discharges (WCDs), maximum most probably discharges, and average most probable discharges (as defined in the C.F.R.) for multiple transit types. When developing sector-specific scenarios, these predefined risks for the regions can be helpful for identifying unique regional concerns and for guiding prioritization for scenario development.

Table B.2 provides a summary of the sector-specific oil spill risk dimensions discussed in each ACP, except for Sector Buffalo. Note that ACPs are updated every five years, so the most current ACP should be referenced whenever possible.

Table B.2. Sector-Specific Oil Spill Risk Dimensions

Sector	Sector Sault Sainte Marie	Sector Lake Michigan	Sector Detroit
ACP name	Northern Michigan Area Contingency Plan (2021)	Sector Lake Michigan Area Contingency Plan (2019c)	Northwest Ohio & Southeast Michigan Area Contingency Plan (2019)
Worst Case Discharge, Vessel	"The largest tank vessel has capacity of 75,298 bbls (3,162,516 gallons) of petroleum or chemical cargoes. As a result, the most likely location for a worst	"Various foreign flagged tank ships make trips to Burns Harbor,	"Largest tank vessel is the M/V ALGONORTH with their Vessel Response Plan (VRP) listing a worst case

Sector	Sector Sault Sainte Marie	Sector Lake Michigan	Sector Detroit
	case discharge (WCD) by vessel is in the vicinity of the St. Mary's River. This WCD scenario was ranked #1 by the NMAC during the 2013 Risk Assessment for a vessel. If a WCD spill occurred in this area, major geographic and logistical difficulties would hamper the response and cleanup operations. *The product would most likely be a light to intermediate grade oil.* Similar tankers transfer fuel at the U.S. Oil facility, ranked #2 during the 2013 Risk Assessment, and Nobel Petro facility, ranked #3, mentioned below and thus would also be potential locations for a WCD by a vessel. As of May 1999, no tank vessel traffic is allowed to transit on the U.S. side of Lake Superior due to failure to meet OPA- 90 response equipment requirements. However, per Canadian Coast Guard information, each year approximately 25 oil tankers transit across Lake Superior on the Canadian side to Thunder Bay, Ontario."	IN, the largest with a capacity of 149,000 bbls (6,258,000 gals) of petroleum products. A tank vessel sinking in Lake Michigan with a total loss of cargo could impact large sections of the shoreline."	discharge at 134,115 bbls. Other worst case vessels in the area include the following: b. M/V ALGOSEA WCD 119,396 bbls c. M/V ALGOCANADA WCD 78K bbls d. M/V ESTA DESGAGNES WCD 68k bbls e. M/V DARA DESGAGNES WCD 68k bbls; ESI of spill location - Humbug Marsh south of Trenton, MI and St. Clair Flats are the worst case scenario for any of these WCD and worst case scenario would be in cold weather/ice conditions"
Worst Case Discharge, Facility	"The largest sources of a worst-case discharge into navigable waters is US Venture US Oil facility located in Rogers City, MI adjacent to Lake Huron. The facility has one main storage tank. WCD is estimated to be 80,000 bbls (3,360,000 gallons). Enbridge Energy also has a pipeline that crosses the Straits of Mackinac five miles west of the Mackinac Bridge. According to 49 C.F.R. § 194.105, a WCD from Line 5 in the Straits of Mackinac is estimated to be 6,658 bbls (279,636 gallons) provided the Emergency Flow Restricting Device (EFRD) closes properly after indication of a release. This is based on an assumed 13 minute time frame for event recognition and remote operated EFRD closures, assuming full line fill volume in the two lines. Products in Line 5 are primarily light and medium crude oil and natural gas liquids. This scenario was ranked #1 overall during the 2013 NMAC Risk Assessment based on Severity, Probability and Impact."	"A worst case discharge at the BP facility in Whiting, IN could result in 577,900 bbls (24,271,800 gals) of Group II oil discharging into the Lake George Barge Canal, thence into Lake Michigan."	No description in ACP.

Sector	Sector Sault Sainte Marie	Sector Lake Michigan	Sector Detroit
Worst Case Discharge, Pipeline	"6658 bbls (279,636 gallons) from Enbridge Line 5 in the Straits of Mackinac."	"A WCD would be complete line segment loss of up to 9,996 bbls (420,000 gals) that could impact Lake Michigan by discharging through the Indiana Harbor Canal."	"Enbridge, Inc. 30" crude oil pipeline that crosses the St. Clair River between Marysville, MI and Froomfield, Ontario. Pipeline carries High Sweet Clearbrook Crude Oil (usual product). Enbridge, through an agreement with State of Michigan, no longer delivers Type V (nonfloating) oils through pipelines. This pipeline will eventually be upgraded to a 36" line in the near future, which will also increase the WCD size. a. Spill size - 5000 bbls/210,000 gals; 2. ESI of spill location–St Clair River & St Clair Flats 3.Oil Group spilled - Type III"
Worst Case Discharge, Rail	"2000 bbls (84,000 gallons) from CN Rail across UP AOR"	"Many trains pass near Lake Michigan with multiple tank cars carrying heavy oil with a capacity of 34,000 gls in each car. A derailment or accident involving a unit train with up to 1 million gallons of product could result in a significant spill of heavy oil in a tributary or drainage ditch leading to Lake Michigan."	No description in ACP.
Worst Case Discharge, Roads	"357 bbls (15,000 gallons) tanker truck"	No description in ACP.	No description in ACP.
Maximum Most Probable Discharge, Vessel (up to 2,500 barrels of oil for vessels with an oil cargo capacity equal to or greater than 25,000 barrels; or 10 percent of the vessels oil cargo capacity for vessels with a capacity of less than 25,000 barrels)	"The maximum most probable discharge based on the largest recent spill recorded would be a 50,000-gallon (vessel use) spill of diesel or bunker-C fuel due to a vessel collision or allision. However, the maximum "most probable", with the emphasis on "most probable" to be a spill of <8000 gallons from a vessel release due to collision or allision."	No description in ACP.	"Freighter transiting the area with allision/collision/grounding resulting in contents of fuel tanks being discharged; 75K to 100K gals of No. 2 or No. 6 Fuel Oil."

Sector	Sector Sault Sainte Marie	Sector Lake Michigan	Sector Detroit
Maximum Most Probable Discharge, Non-Vessel (lesser of 1,200 barrrels or 10 percent volume of WCD)	No description in ACP.	No description in ACP.	“Maximum Most Probable Discharge (non-Vessel): overturned tank truck (approximately 8,000 gallons) on roadway/bridge over a navigable waterway leading to discharge of product. • St. Clair Tunnel (Port Huron to Sarnia): Majority of HAZMAT in AOR transit by railway and may lead to Overturned rail cars w/ flammable or dangerous goods. Percentage of dangerous goods by class transported through the tunnel over a 12 month period • • 2.1 Flammable gases: 33% • • 2.2 Non-flammable, Non-toxic gases: 1% • • 2.3 Toxic gases: 3% • • 3 Flammable Liquids: 27% • • 4 Flammable Solids and Reactive Solids/Liquids: 2% • • 5 Oxidizers and Organic Peroxides: 3% • • 6 Poisons Materials and Infectious Substances: 2% • • 8 Corrosive Materials: 22% • • 9 Miscellaneous Products, Substances or Organisms: 7% “
Average Most Probable Discharge, Non-Vessel (the lesser of 50 barrels or 1 percent of the volume of the worst case discharge)	“<250 gallons and is based on the size of the average spill in the COTP area from 2018–2022. Likely sources of this type of spill are from a (1) pleasure craft bilge pumping, a spill while fueling and or the vessel sinking; (2) a commercial vessel’s shaft seal leak, spill while bunkering, or a bilge discharge; (3) a spill from a land-based vehicle accident that enters a tributary or navigable body of water. *The product would most likely be a lighter end petroleum distillate product such as gasoline, diesel oil or lube oil.”*	No description in ACP.	“Most Probable Discharge: Release from bilge, fueling spill, or seal leak from commercial or recreational vessel; less than 500 gallons.”
Average Most Probable Discharge, Vessel (50 barrels of oil from the vessel)	No description in ACP.	No description in ACP.	No description in ACP.

SOURCES: U.S. Coast Guard, 2019; U.S. Coast Guard, Northwest Ohio & Southeast Michigan Area Contingency Plan, Sector Detroit, May 13, 2019; U.S. Coast Guard, Northern Michigan Area Contingency Plan, 2021.

Appendix C. Structured Discussion Protocol

The following is the interview protocol that we used during the structured SME discussions for this project.

Great Lakes Oil-Spill Response Capabilities, Capacity, and Gaps Structured Discussion Protocol

Contents

1. Project Overview
2. Consent Script
3. Questions/Topics of Interest

Project Overview

The Homeland Security Operational Analysis Center (HSOAC) is examining the ability of government agencies and the private sector to respond to oil spills in the Great Lakes, where spills could pose a threat to the region's ecology, economy, and public health. This study will describe capabilities and capacities related to Area Contingency Plans (ACPs) and identify any gaps in those abilities. It will also provide pot on how those gaps can be addressed in ways that are actionable for the Great Lakes Oil Spill Center of Expertise (GLCOE), Coast Guard District Nine and Atlantic Area, and other stakeholders.

Consent Script

Before we get started, we wanted to make sure that you understood the purpose of this research project and how anything we discuss today will be used.

This project is sponsored by Coast Guard Headquarters, Office of Marine Environmental Response Policy (CG-MER-3), and is being conducted by the Homeland Security Operational Analysis Center (HSOAC), which is a federally funded research and development center (FFRDC) for the Department of Homeland Security (DHS) operated by the RAND Corporation.

The purpose of this project is to help the Coast Guard understand

1. the scope of current risks of an oil spill
2. the roles and capacity of partners performing oil-spill response
3. an analysis of gaps and mitigation strategies around oil-spill response.

We will be asking you questions about how your organization contributes to oil-spill response in the Great Lakes region and about any gaps or mitigations to oil-spill response your organization has identified. Participation in this discussion is voluntary. If you do not wish to answer any of these questions, you do not have to, and you may leave at any time.

The results of our discussion may be included in a publicly available report. While we will not use your name in this report, we may attribute your comments to your organization, so it is possible that what you say here today will not remain anonymous.

Do you have any questions and are you OK with proceeding?

Contact information for further questions:

- Kate Anania, kanania@rand.org, 703-413-1100 x 5364 (office)
- Aaron Davenport, aarond@rand.org

Human Subject Protection Committee contact information:

If you have questions about your rights as a research participant or need to report a research-related injury or concern, you can contact RAND's Human Subjects Protection Committee toll-free at (866) 697-5620 or by emailing hspcinfo@rand.org. If possible, when you contact the committee, please reference Study #2022-N0545.

Questions/Topics of Interest

HSOAC seeks stakeholder and SME input regarding the following:

Topics of Interest

Scale and scope of the problem. We want to characterize the flow of oil in the Great Lakes and their immediate watershed: roughly what quantities of oil are moved by pipeline, by vessel, and by other means in and around the Great Lakes? Are there any indications of potential large-scale changes in the next couple of decades?

Oil-spill response capabilities and capacities. We want information about government agencies and private companies on both sides of the U.S.-Canadian border: different types of cleanup capabilities that are being considered, recognizing that capabilities and capacities may be affected by specific environmental, seasonal, and other factors and acknowledging that the field of fresh water oil-spill response science is less mature than that of salt water.

Stakeholder and SME questions:

1. What are the responsibilities of your organization for response operations?
   a. Do you have a response plan or standard operating procedure? When was it last updated?
   b. What equipment is required for response operations? Where is the equipment located?
   c. Are there requirements for response capacity and timelines?

d. What skillsets are required for response operations?
e. What training is required of your response personnel (focus and frequency; individual, unit, interagency)? Do personnel have certifications?
f. How often does your organization/unit engage in exercises? Are other organizations involved? Which ones?
    i. Are there lessons learned/documentation from these exercises? (Please share if possible.)

2. What other organizations do you coordinate with for spill response?
    a. How are response operations coordinated?
    b. What kinds of agreements do you have in place?
    c. What are the strengths of this coordination?
    d. What are the weaknesses of this coordination?

3. Do you have information about the physical locations of capabilities and how quickly they can be deployed to where they may be needed (both capabilities owned by your organization and others)? Who owns/is responsible for these capabilities?

4. Are you aware of or does your organization consider any specific oil-spill scenarios with regard to the physical environment, maps and models, as well as any unique regional hydrology and ice conditions?
    a. Are there specific locations, spill quantities, oil type(s), etc. that are of particular concern?
    b. What scenarios would have the greatest negative impacts?
    c. What conditions mean responsible organizations could respond to a spill effectively and efficiently?

5. How can the use of emerging technologies or tactical response approaches increase the efficacy of response activities?
    a. What technologies/tactical response approaches are currently most effective?
    b. What opportunities (technology, response approaches etc.) would you like to see improved for Great Lakes oil-spill response?

6. Are there other approaches that include changes to statutory authorities, improved training, intensified coordination among multiple U.S. and Canadian agencies and the private sector?

7. Are you aware of any relevant scientific oil-spill research that is fresh water-specific or ice-specific? This includes any enhanced characterization of oil distribution over time through the use of sensors and spill modeling.

8. Are there rivers or wetlands (or other areas not on or under the Great Lakes proper) where a spill would impact the lakes or that the NCOE should be concerned about?

Additional Items of Interest

1. Any available after-action reports from Great Lakes oil-spill response exercises and/or any oil-spill response exercises in icy conditions

2. Organizational planning meeting minutes regarding oil-spill response capabilities or other documentation that elucidates capability gaps in spill response
3. Spill response SOPs or plans
4. Any documentation that indicates types and conveyance of HAZMAT through the Great Lakes region (ship, pipeline, tank trunk, tank car[rail])
5. Any documentation that indicates any point sources of HAZMAT
6. Historical oil spill data (location, type, amount, etc.).

Abbreviations

ACP	Area Contingency Plans
AIS	automatic identification system
AOR	area of responsibility
BE	burn efficiency
BOEM	Bureau of Ocean Energy Management
CANUSLAK	Great Lakes annex of the International Convention on Oil Pollution Prevention Response and Cooperation of 1990
CCG	Canadian Coast Guard
C.F.R.	Code of Federal Regulations
DHS	U.S. Department of Homeland Security
ECRC	Eastern Canada Response Corporation
EPA	Environmental Protection Agency
ERMA	Environmental Response Management Application
ESI	Environmental Sensitivity Index
FOSC	Federal On Scene Coordinator
GRS	Geographic Response Strategies
HSOAC	Homeland Security Operational Analysis Center
ISB	in-situ burn
NASA	National Aeronautics and Space Administration
NCP	National Oil and Hazardous Substances Pollution Contingency Plan
NOAA	National Oceanic and Atmospheric Administration
NSF	National Strike Force
OSRO	Oil Spill Response Organization
RCP	Regional Contingency Plan
SME	subject-matter expert
TTP	tactics, techniques, and procedures
UUV	unmanned underwater vehicles
WCD	worst case discharges
USCG	U.S. Coast Guard

References

33 U.S.C. § 1251, Title 40, Chapter I, Subchapter D, Part 112, Oil Pollution Prevention, as amended 2002. As of December 5, 2023: https://www.ecfr.gov/current/title-40/chapter-I/subchapter-D/part-112

33 U.S.C. § 1320, Federal Water Pollution Control Act, Oil and Hazardous Substance Liability, 2002. As of December 5, 2023: https://www.epa.gov/sites/default/files/2017-08/documents/federal-water-pollution-control-act-508full.pdf

33 U.S.C. § 2701, Chapter 40, Oil Pollution, Definitions. As of December 5, 2023: https://uscode.house.gov/view.xhtml?path=/prelim@title33/chapter40&edition=prelim

Annex 1, Great Lakes Operational Supplement to the Joint Marine Pollution Contingency Plan, Canadian Coast Guard and United States Coast Guard, 2022.

Arctic Council, Emergency Prevention, Preparedness and Response, "Guide to Oil Spill Response in Snow and Ice Conditions in the Arctic," 2015.

Bartolai, A., L. He, A. Hurst, L. Mortsch, R. Paehlke, and D. Scavia, "Climate Change as a Driver of Change in the Great Lakes St. Lawrence River Basin," *Journal Of Great Lakes Research*, 2015

Bejarano, Adriana C., "Critical Review and Analysis of Aquatic Toxicity Data on Oil Spill Dispersants," *Environmental Toxicology and Chemistry*, Vol. 37, No. 12, 2018.

Black, T. A., M. S. White, J. M. Blais, B. Hollebone, D. M. Orihel, V. P. Palace, J. L. Rodriguez-Gil, and M. L. Hanson, "Surface Oil is the Primary Driver of Macroinvertebrate Impacts Following Spills of Diluted Bitumen in Freshwater," *Environmental Pollution*, Vol. 290, 2021. December 5, 2023: https://doi.org/10.1016/j.envpol.2021.117929

Bloomberg, "Enbridge," company profile, undated. As of December 5, 2023: https://www.bloomberg.com/profile/company/ENB:CN#xj4y7vzkg

Boak, Jeremy, and Robert Kleinberg, "4–Shale Gas, Tight Oil, Shale Oil and Hydraulic Fracturing," Future Energy (Third Edition), 2020. As of June 14, 2023: https://www.sciencedirect.com/science/article/abs/pii/B9780081028865000049

BP, "Our Pipelines," webpage, undated. As of December 5, 2023: https://www.bp.com/en_us/united-states/home/products-and-services/pipelines/our-pipelines.html#accordion_crude

Buist, Ian, James McCourt, Steve Potter, Sy Ross, and Ken Trudel, "In Situ Burning," Pure Applied Chemistry, Vol. 71, No. 1, 1999.

Canada–United States Joint Contingency Plan, Canadian Coast Guard and United States Coast Guard, August 3, 2017.

Canadian Coast Guard Environmental Response, *Great Lakes Joint Response Team Harvest Spirit Grounding After Action Report: 02 Dec–04 Dec 2020*, provided to authors, June 1, 2021a.

Canadian Coast Guard Environmental Response, *CANUSLAK 2021 Exercise After Action Report 18 August 2021*, provided to authors, September 22, 2021b.

Canadian Coast Guard Environmental Response, *CANUSLAK 2022 Exercise After Action Report: 17 May 2022*, provided to authors, July 7, 2022a.

Canadian Coast Guard Environmental Response, *Great Lakes Joint Response Team Barge PML 9000 Grounding After Action Report 04 May–07 May 2022*, provided to authors, August 25, 2022b.

Canadian Coast Guard Environmental Response, *Great Lakes Joint Response Team Tecumseh Grounded Vessel After Action Report: 18 June–22 June 2022*, provided to authors, September 21, 2022c.

Canadian Coast Guard Environmental Response, *Great Lakes Joint Response Team M/V SAM LAUD Grounding After Action Report: 08 Oct–09 Oct 2022*, provided to authors, January 18, 2023.

CANUSLAK Joint Response Team, Annual Meeting Minutes, March 14, 2022

Cedre, "Dispersants Approved by Cedre For Oil Spill Response," April 2023. As of December 5, 2023: https://wwz.cedre.fr/en/content/download/3069/file/04_2023_internet_dispersants_Freshwater_gb.pdf

CCG—*See* Canadian Coast Guard.

C.F.R.—*See* Code of Federal Regulations.

Changnon, S., "Changes in Climate and Levels of Lake Michigan: Shoreline Impacts at Chicago," *Climatic Change,* Vol. 23, No. 3, 1993

Channell, Kim, John DelPizzo, Laura Briley, Richard Rood, Jenna Jorns, and Kate Hutchens, "Lake Ontario: A Summary of Anticipated Future Climate Conditions," Great Lakes Integrated Sciences and Assessments, September 2022. As of December 5, 2023: https://glisa.umich.edu/wp-content/uploads/2022/09/Superior_Prospective_Report_2022.pdf

Coast Guard District 9, "CGD 9 Input for Reply to ELPC Brief," September 18, 2019.

Congressional Research Service, *U.S. Rail Transportation of Crude Oil: Background and Issues for Congress*, R43390, December 4, 2014. As of December 5, 2023: https://www.everycrsreport.com/reports/R43390.html

Daisy, Naznin Sultana, Mohammad Hesam Hafezi, Lei Liu, and Kenneth Lee, "A Comprehensive Review of Canadian Marine Oil Spill Response System Through the Lens of Decanting Regulations and Practices," *Journal of Marine Science and Engineering*, Vol. 10, No. 9, 2022.

Demirel, F. Bugra, and Ersan Basar, "Applications of a New Product To Decrease the Smoke Plume Opacity of In-Situ Burning of Oil Spills," *Fresenius Environmental Bulletin*, Vol. 25, No. 11, 2016.DHS—*See* U.S. Department of Homeland Security.

Eastern Canada Response Corporation, "About ECRC," webpage, undated. As of December 5, 2023:
https://www.ecrc-simec.ca/en/about/ecrc/

Elastec, "Weir Skimmers," webpage, undated. As of December 5, 2023: https://www.elastec.com/products/oil-spill-skimmers/weir-oil-skimmers/

Enbridge, "Line 5 Through the Bad River Reservation," webpage, undated a. As of December 5, 2023:
https://www.enbridge.com/projects-and-infrastructure/public-awareness/line-5-wisconsin-segment-relocation-project/line-5-through-the-bad-river-reservation

Enbridge, "Liquids Pipelines," webpage, undated b. As of July 10, 2023: https://www.enbridge.com/about-us/liquids-pipelines

Enbridge, *The Mainline Pipeline System*, fact sheet, undated c. As of December 5, 2023: https://www.enbridge.com/~/media/Enb/Documents/Factsheets/FS_ENB_Mainline_system.pdf

Enbridge, *Investing in Michigan's Future*, March 2022. As of December 5, 2023: https://www.enbridge.com/-/media/Enb/Documents/Projects/line5/Investing-in-Michigans-future-FINAL.pdf?rev=c3438e4d718b49c0b2ae39115e02e59d&hash=9A34FD9941D35EFF59483D7AC06ADB4F

Enbridge, *Enbridge's Energy Infrastructure Assets*, April 19, 2023. As of June 13, 2023: https://www.enbridge.com/~/media/Enb/Documents/Factsheets/FS_EnergyInfrastructureAssets.pdf?la=en

Environmental Law and Policy Center and National Wildlife Federation v. United States Coast Guard, 447 F. Supp. 3d 602, (E.D. Mich. 2020).

Environmental Protection Agency, "Discharge of Oil Regulation Overview," webpage, Environmental Protection Agency Archive, undated. As of December 5, 2023: https://archive.epa.gov/emergencies/content/lawsregs/web/html/sheenovr.html

Environmental Protection Agency, "Types of Oil," webpage, January 24, 2022a. As of October 31, 2022: https://www.epa.gov/emergency-response/types-oil

Environmental Protection Agency, "Climate Change Indicators: Great Lakes Water Levels and Temperatures," webpage, August 2, 2022b. As of June 25, 2023: https://www.epa.gov/climate-indicators/great-lakes

Environmental Protection Agency, "Skimmers," updated September 2022c. As of April 5, 2023: https://www.epa.gov/emergency-response/skimmers

Environmental Protection Agency, "Booms," October 2022d. As of April 2023: https://www.epa.gov/emergency-response/booms

Environmental Protection Agency, "Dispersants," updated November 2022e. As of April 2023: https://www.epa.gov/emergency-response/dispersants

Environmental Protection Agency, "EPA's Response Techniques," webpage, updated December 2022f. As of April 2023: https://www.epa.gov/emergency-response/epas-response-techniques

EPA—*See* Environmental Protection Agency.

ERMA—*See* National Oceanic and Atmospheric Administration, Great Lakes Environmental Response Management Application.

Executive Order 7521, "Use of Vessels for Ice-Breaking Operations in Channels and Harbors," December 21, 1936.

Faksness, Liv-Guri, Frode Leirvik, Ingrid C. Taban, Frode Engen, Hans V. Jensen, Jan Willie Holbu, Hilde Dolva, and Magne Båtveit, "Offshore Field Experiments with In-situ Burning of Oil: Emissions and Burn Efficiency," *Environmental Research*, Vol. 205, 2022.

Frittelli, John, *Shipping on the Great Lakes and St. Lawrence Seaway: An Update*, Congressional Research Service, R47550, May 9, 2023. As of December 5, 2023: https://sgp.fas.org/crs/misc/R47550.pdf

GAO—*See* U.S. Government Accountability Office.

Geographic Response Strategy, "Algoma, WI: WLM17-ALG," USGS, undated a.

Geographic Response Strategy, "Baileys Harbor, WI: WLM24-BAL," USGS, undated b.

Geographic Response Strategy, "Burns Harbor, IN: SLM1-BUR," USGS, undated c.

Geographic Response Strategy, "Chicago, IL: SLM7-CHI," USGS, undated d.

Geographic Response Strategy, "East Chicago, IN: SLM4-ECH," USGS, undated e.

Geographic Response Strategy, "Ephraim Bay, WI: WLM32-EPH," USGS, undated f.

Geographic Response Strategy, "Escanaba/Little Bay DeNoc: NLM5-ESC," USGS, undated g.

Geographic Response Strategy, "Evanston, IL: SLM8-EVA," USGS, undated h.

Geographic Response Strategy, "Frankfort, MI: ESI# ELM1-FRA," USGS, undated i.

Geographic Response Strategy, "Grand Haven/Port Sheldon: ELM09-GRH," USGS, undated j.

Geographic Response Strategy, "Green Bay, WI: WLM42-GBS," USGS, undated k.

Geographic Response Strategy, "Holland and Saugatuk: ELM11-HOL," USGS, undated l.

Geographic Response Strategy, "Jacksonport, WI: WLM22-JAX," USGS, undated m.

Geographic Response Strategy, "Kenosha, WI: WLM1-KEN," USGS, undated n.

Geographic Response Strategy, "Little Sable Pt./Muskegon/White Lake: ELM07-MUS," USGS, undated o.

Geographic Response Strategy, "Little Sturgeon Bay, WI: WLM38-LSB," USGS, undated p.

Geographic Response Strategy, "Ludington, MI: ESI# - ELM3-LUD," USGS, undated q.

Geographic Response Strategy, "Manistee, MI: ESI# - ELM2-MAN," USGS, undated r.

Geographic Response Strategy, "Manitowoc, WI: WLM13-MAN," USGS, undated s.

Geographic Response Strategy, "Marinette, WI: WLM51-MAR," USGS, undated t.

Geographic Response Strategy, "Michigan City/New Buffalo: ELM23-MIC," USGS, undated u.

Geographic Response Strategy, "Milwaukee, WI: WLM5-MKE," USGS, undated v.

Geographic Response Strategy, "Port Washington, WI: WLM8-PWA," USGS, undated w.

Geographic Response Strategy, "Racine, WI: WLM2-RAC," USGS, undated x.

Geographic Response Strategy, "Sector Detroit GRS Div 3: Bay City," USGS, undated y.

Geographic Response Strategy, "Sector Detroit GRS Div 4: Saginaw Bay," USGS, undated z.

Geographic Response Strategy, "Sector Detroit Lake Huron Division 1," USGS, undated aa.

Geographic Response Strategy, "Sector Detroit Saginaw Bay Division 2," USGS, undated bb.

Geographic Response Strategy, "Sheboygan, WI: WLM10-SHE," USGS, undated cc.

Geographic Response Strategy, "Sister Bay, WI: WLM24-SIS," USGS, undated dd.

Geographic Response Strategy, "Sleeping Bear Dunes: NLM64-SBD," USGS, undated ee.

Geographic Response Strategy, "South Haven: ELM14-SHA," USGS, undated ff.

Geographic Response Strategy, "ST Joseph: ELM17-STJ," USGS, undated gg.

Geographic Response Strategy, "Sturgeon, Bay, WI: WLM20-STB," USGS, undated hh.

Geographic Response Strategy, "Two Rivers, WI: WLM14-TWO," USGS, undated ii.

Geographic Response Strategy, "Washington Island, WI: WLM27-WAS," USGS, undated jj.

Geographic Response Strategy, "Waukegan, IL: WLM11-WAK," USGS, undated kk.

Giachino, Annette, "Enbridge Provides Update on Tunnel Project," *Upper Michigan's Source*, February 20, 2023. As of December 5, 2023: https://www.uppermichiganssource.com/2023/02/20/enbridge-provides-update-tunnel-project/

Gibb, Terry, "Lakes Appreciation Month: The Great Lakes Facts and Features," Michigan State University Extension, July 20, 2015. As of December 5, 2023: https://www.canr.msu.edu/news/lakes_appreciation_month_the_great_lakes_facts_and_features#:~:text=35%20million%20people%20live%20in%20the%20Great%20Lakes,used%20daily%20in%20municipal%2C%20agricultural%20and%20industrial%20activities

Great Lakes Carriers Association, "Great Lakes Icebreaking," position paper, June 16, 2020. As of May 5, 2023: https://lcaships.com/http-www-lcaships-com-wp-content-uploads-2020-06-icebreaking-resources-10-june-2020-final-pdf/

Great Lakes Commission, *Issues and Trends Surrounding the Movement of Crude Oil in the Great Lakes–St. Lawrence River Region*, February 2015. As of December 4, 2023: www.glc.org/wp-content/uploads/Oil-Transportation-Full-Report-2015.pdf

Great Lakes Observing System, Seagull, web application, undated. As of December 5, 2023: https://seagull.glos.org

GRS—*See* Geographic Response Strategy.

The Greenbrier Companies, *28,600 Gallon DOT 117 Tank Car*, fact sheet, January 2021. As of December 5, 2023: https://www.gbrx.com/wp-content/uploads/2021/06/28600-Gallon-DOT-117-Tank-Car.pdf

Hazardous Material Transport Outreach Network, "Crude Oil Transport by Pipeline in the Great Lakes," webpage, undated a. As of December 5, 2023: https://hazmaton.org/pipelines/

Hazardous Material Transport Outreach Network, "Crude Oil Transport by Truck in the Great Lakes," webpage, undated b. As of December 5, 2023: https://hazmaton.org/crude-oil-transport-by-truck-in-the-great-lakes/

HAZMATON—*See* Hazardous Material Transport Outreach Network.

Hosea, Leana, "Why Michigan is Trying to Shut Down Canada's Enbridge Line 5 Pipeline," BBC, January 15, 2023. As of June 28, 2023: https://www.bbc.com/news/world-us-canada-63879493

Hull, Bradley, III, *Status of Infrastructure Related to Crude Oil Transportation in the Great Lakes/St. Lawrence River Region*, May 31, 2017. As of December 5, 2023: https://www.glc.org/wp-content/uploads/Oil-Transp-Infrastructure-Hull-May2017-FinalReport.pdf

ICCOPR—*See* Interagency Coordinating Committee on Oil Pollution Research.

Interagency Coordinating Committee on Oil Pollution Research, *Oil Pollution Research and Technology Plan: Fiscal Years 2022–2027*, December 2021. As of December 5, 2023: https://www.dco.uscg.mil/Portals/9/CG-5R/ICCOPR/Files/2022_2027%20ICCOPR%20ORTP%20Plan.pdf?ver=XcMuMlBf8q2RymJNKcwzUg%3D%3D&timestamp=1642090228492

International Institute for Sustainable Development, Experimental Lakes Area, "Who We Are," webpage, undated. As of April 24 2023: https://www.iisd.org/ela/about/who-we-are/

International Joint Commission, "Role of the IJC," webpage, undated. As of December 5, 2023: https://www.ijc.org/en/who/role

International Joint Commission, Great Lakes Water Quality Board, 2021 Great Lakes Regional Poll, November 2021a. As of December 5, 2023: https://ijc.org/en/wqb/great-lakes-poll

International Joint Commission, Great Lakes Water Quality Board, *2021 Great Lakes Regional Poll Online Poll Final Report*, November 2021b. As of December 5, 2023: https://ijc.org/sites/default/files/WQB_GreatLakesRegionalPoll_OnlinePollReport_2021.pdf

International Joint Commission, Great Lakes Water Quality Board, 2021 Great Lakes Regional Poll, Key Findings, November 2021c. As of December 5, 2023: https://ijc.org/sites/default/files/2021_WQB-Great-Lakes-poll_one-pager_US-CAN.pdf

International Tanker Owners Pollution Federation Limited, *Use of Booms in Oil Pollution Response*, Technical Information Paper 3, 2011.

International Tanker Owners Pollution Federation Limited, *Fate of Marine Oil Spills*, Technical Information Paper 2, 2014a.

International Tanker Owners Pollution Federation Limited, *Use of Skimmers in Oil Pollution Response*, Technical Information Paper 5, 2014b. As of July 12, 2023: https://www.itopf.org/fileadmin/uploads/itopf/data/Documents/TIPS_TAPS_new/TIP_5_Use_of_Skimmers_in_Oil_Pollution_Response.pdf

Kompoltowicz, Keith, "Understanding Great Lakes Water Level Fluctuations and the Need for Coastal Resiliency," *Proceedings*, Vol. 79, No. 3, Winter 2022.

Krouse, Peter, "Cuyahoga County Has More Than 300 Orphan Oil and Gas Wells That May Need Plugging," Cleveland.com, March 24, 2022. As of December 10, 2023: https://www.cleveland.com/news/2022/03/cuyahoga-county-has-more-than-300-orphan-oil-and-gas-wells-that-may-need-plugging.html

Lewis, Chelsey, "Huge Waves Hit Lake Michigan's Eastern Shore, After Lake Superior Waves Set Record in October," Milwaukee Journal Sentinel, December 7, 2017. As of June 2, 2023: https://www.jsonline.com/story/weather/2017/12/07/huge-waves-hit-lake-michigans-eastern-shore-after-lake-superior-waves-set-record-october/930623001/

LimnoTech, *Impacts of Unrefined Liquid Hydrocarbons on Water Quality and Aquatic Ecosystems of the Great Lakes Basin*, January 2018.

Marchand, D., M. Sanderson, D. Howe, and C. Alpaugh, "Climatic Change and Great Lakes Levels The Impact On Shipping," *Climatic Change*, Vol. 12, No. 2, 1988.

McCarten, James, "Enbridge 'Must Cease' Line 5 Operations on Bad River Band Territory by June 2026: Judge," The Canadian Press, June 27, 2023. As of December 5, 2023: https://globalnews.ca/news/9796449/enbridge-line-five-judge-wisconsin-bad-river/

McWhirter, Sheri, "Wisconsin Line 5 Shutdown Order May Disrupt Pipeline Fight in Michigan," Michigan Live, June 21, 2023. As of December 5, 2023: https://www.mlive.com/public-interest/2023/06/wisconsin-line-5-shutdown-order-may-disrupt-pipeline-fight-in-michigan.html

Michigan Tech, "Definitions of Lake Ice Terms," webpage, undated. As of June 12, 2023: geo.mtu.edu/great_lakes/icegroup/ice_terms_jake.html

NASA—*See* National Aeronautics and Space Administration.

National Academies of Sciences, Engineering, and Medicine, "Responding to Oil Spills in the U.S. Arctic Marine Environment," National Academies Press, 2014. As of December 5, 2023: https://doi.org/10.17226/18625

National Aeronautics and Space Administration, Earth Science Applied Sciences, "Supporting Recovery Efforts After Oil Spills," webpage, undated. As of December 5, 2023: https://appliedsciences.nasa.gov/what-we-do/disasters/oil-spills

National Oceanic and Atmospheric Administration, "Storm Events Database," webpage, undated. As of June 2, 2023: https://www.ncdc.noaa.gov/stormevents/

National Oceanic and Atmospheric Administration, Great Lakes Environmental Research Laboratory, "Great Lakes Water Level Observations," undated a. As of June 2, 2023: https://www.glerl.noaa.gov/data/wlevels/#observations

National Oceanic and Atmospheric Administration, Great Lakes Environmental Research Laboratory, "Ice Cover," brochure, undated b. As of December 5, 2023: https://www.glerl.noaa.gov/pubs/brochures/ice.pdf

National Oceanic and Atmospheric Administration, Great Lakes Environmental Research Laboratory, "Water Levels," brochure, undated c. As of June 25, 2023: https://www.glerl.noaa.gov/pubs/brochures/lakelevels/lakelevels.pdf

National Oceanic and Atmospheric Administration, Great Lakes Environmental Research Laboratory, "About Our Great Lakes, Lake by Lake Profiles," June 2004. As of December 5, 2023: https://www2.census.gov/library/publications/2010/compendia/statab/130ed/tables/11s0356.pdf

National Oceanic and Atmospheric Administration, Great Lakes Environmental Research Laboratory, "GLCFS Modeled NOWCAST Wave Heights vs Observed Wave Heights," webpage, 2017. As of June 2, 2023: https://www.jsonline.com/story/weather/2017/12/07/huge-waves-hit-lake-michigans-eastern-shore-after-lake-superior-waves-set-record-october/930623001/

National Oceanic and Atmospheric Administration, Great Lakes Environmental Research Laboratory, "Great Lakes Ice Cover," webpage, updated March 2023. As of April 5, 2023: https://www.glerl.noaa.gov/data/ice/#overview

National Oceanic and Atmospheric Administration, Great Lakes Environmental Response Management Application, accessed June 1, 2023. As of December 5, 2023: https://erma.noaa.gov/greatlakes#

National Oceanic and Atmospheric Administration, National Ocean Service, "What is Eutrophication?" webpage, undated a. As of December 5, 2023: https://oceanservice.noaa.gov/facts/eutrophication.html

National Oceanic and Atmospheric Administration, National Ocean Service, "Now Does NOAA Help Clean Up Oil and Chemical Spills?" webpage, undated b. As of December 5, 2023: https://oceanservice.noaa.gov/facts/spills-cleanup.html

National Oceanic and Atmospheric Administration, Office for Coastal Management, *2022 Marine Economy Report*, undated. As of December 5, 2023: https://coast.noaa.gov/data/digitalcoast/pdf/econ-report-regional-state.pdf

National Oceanic and Atmospheric Administration, Office for Coastal Management, "Great Lakes," 2023. As of October 4, 2023: https://www.coast.noaa.gov/states/fast-facts/great-lakes.html

National Oceanic and Atmospheric Administration, Office of Response and Restoration, "Chemistry of an Oil Spill," webpage, August 21, 2012. As of December 4, 2023: https://response.restoration.noaa.gov/about/media/chemistry-oil-spill.html

National Oceanic and Atmospheric Administration, Office of Response and Restoration, "How Is an Oil Spill in a River Different Than One in the Ocean?" webpage, July 31, 2015. As of December 5, 2023: https://response.restoration.noaa.gov/about/media/how-oil-spill-river-different-one-ocean.html

National Oceanic and Atmospheric Administration, Office of Response and Restoration, "What Have We Learned About Using Dispersants During the Next Big Oil Spill?" April 2015, updated November 2022. As of April 2023: https://response.restoration.noaa.gov/about/media/what-have-we-learned-about-using-dispersants-during-next-big-oil-spill.html

National Oceanic and Atmospheric Administration, Office of Response and Restoration, "Residues from In Situ Burning of Oil on Water," February 2019. As of December 5, 2023: https://response.restoration.noaa.gov/oil-and-chemical-spills/oil-spills/resources/residues-in-situ-burning-oil-water.html

National Oceanic and Atmospheric Administration, Office of Response and Restoration, "Great Lakes Coast Guard Oil Spill Center of Expertise Grand Opening," webpage, August 29, 2022. As of December 5, 2023: https://response.restoration.noaa.gov/great-lakes-coast-guard-oil-spill-center-expertise-grand-opening

National Oceanic and Atmospheric Administration and Bureau of Ocean Energy Management, "AISVesselTracks2018GreatLakes," database, undated. As of May 1, 2023: www.marinecadastre.gov/data

Nelson, Dawn, Heather Elmer, Rebecca Held, Danielle Forsythe, and Shauna Casey, "Laurentian Great Lakes Basin Climate Change Adaptation," NOAA Technical Memorandum GLERL-153, March 2011.

NOAA—*See* National Oceanic and Atmospheric Administration.

North Atlantic Treaty Organization, "AIS (Automatic Identification System) overview," webpage, 2021. As of December 5, 2023: https://shipping.nato.int/nsc/operations/news/2021/ais-automatic-identification-system-overview

Nuka Research and Planning Group, *Inventory and Assessment of Marine Oil Spill Response Resources in Massachusetts and New England States*, December 2009. As of April 7, 2023: https://www.mass.gov/doc/inventory-assessment-of-marine-oil-spill-response-resources-in-massachusetts-new-england-states/download

Ohio Department of Natural Resources, "Orphan Well Program," webpage, undated. As of December 5, 2023: https://ohiodnr.gov/business-and-industry/energy-resources/oil-and-gas-wells/orphan-well-program

Pace, LCDR Christopher, U.S. Coast Guard, "Scenarios and ACPs? Straits exercise," email to authors, February 7, 2023;

Popiel, Jerome A., "Great Lakes Environmental Response," *Proceedings*, Vol. 79, No. 3, Winter 2022.

Popiel, Jerome A., "Sector Detroit ACP.pdf," email to K. Anania, C. Pace, C. J. Graham, and A. Davenport, USCG, February 7, 2023.

Proceedings, *The Great Lakes, The Ninth District's Unique Operating Environment,* Vol. 79, No. 3, Winter 2022. As of December 4, 2023: https://www.dco.uscg.mil/Portals/9/DCO%20Documents/Proceedings%20Magazine/Archive/2022/Vol79_No3_Wint22.pdf?ver=vm_cEFKVR5rCZy8iGyUVTQ%3D%3D

Rayburn, Thomas, "Integrating Data, Resources, and Capabilities in the Great Lakes: Impacting Federal, State, Tribal, Local, and Private Partnerships Through the Area Committee," Great Lakes Commission, 2003.

Region 5 Regional Response Team, *In-Situ Burning Annex*, undated. As of May 10, 2023: https://rrt5.org/Portals/0/Appendices/App7_InSituBurningAnnex_v2015.pdf?ver=2017-07-10-155602-000

Revised Statutes of Canada, Canada Oil and Gas Operations Act, Chapter O-7, 1985.

Rodriguez-Gil, Jose Luis, Sawyer Stoyanovich, Mark L. Hanson, Bruce Hollebone, Diane M. Orihel, Vince Palace, Robert Faragher, Fatemeh S. Mirnaghi, Keval Shah, Zeyu Yang, Tyler A. Black, Jeffrey Cederwall, Johanna Mason, Samuel Patterson, Lauren Timlick, Jonathan Y. Séguin, and Jules M. Blais, "Simulating Diluted Bitumen Spills in Boreal Lake Limnocorrals: Part 1–Experimental Design and Responses of Hydrocarbons, Metals, and Water Quality Parameters," *Science of The Total Environment*, Vol. 790, 2021. As of December 5, 2023: https://doi.org/10.1016/j.scitotenv.2021.148537

Statutes of Canada, Canada Shipping Act, Chapter 26, Part 8, Pollution Prevention and Response—Department of Transport and Department of Fisheries and Oceans, 2001.

Statutes of Canada, Marine Liability Act, Chapter 6, 2001.

Steinman, D. B., Final Geological Report, 1958. As of July 10, 2023: https://www.michigan.gov/-/media/Project/Websites/egle/Documents/Programs/OGMD/Catalog/03/Final-Geological-Report.pdf?rev=1b6e73a303e545dcab9381ed00d32153#:~:text=The%20Straits%20of%20Mackinac%20is,is%20located%20about%20mid%2Dstream

Stoyanovich, S., J. R. Rodríguez-Gil, M. L. Hanson, B. P. Hollebone, D. M. Orihel, V. P. Palace, R. Faragher, F. S. Mirnaghi, K. Shah, Z. Yang, and J. M. Blais, "Simulating Diluted Bitumen Spills in Boreal Lake Limnocorrals: Part 2–Factors Affecting the Physical Characteristics and Submergence of Diluted Bitumen," *Science of The Total Environment*, Vol. 790, 2021, As of December 5, 2023: https://doi.org/10.1016/j.scitotenv.2021.148580

Stoyanovich, S., Z. Yang, M. Hanson, B. P. Hollebone, D. M. Orihel, V. Palace, J. R. Rodriguez-Gil, F. Mirnaghi, K. Shah, and J. M. Blais, "Fate of Polycyclic Aromatic Compounds from Diluted Bitumen Spilled into Freshwater Limnocorrals," *Science of The Total Environment*, Vol. 819, 2022. As of December 5, 2023: https://doi.org/10.1016/j.scitotenv.2021.151993

Tip of the Mitt Watershed Council, "Enbridge Energy Line 5," undated. As of December 5, 2023: https://www.watershedcouncil.org/enbridge-energy-line-5.html

Title 14, Section 2, Primary Duties.

Title 33, Chapter I, Subchapter O, Part 154, Facilities Transferring Oil or Hazardous Material in Bulk.

Title 33, Chapter I, Subchapter O, Part 155, Subpart D, Tank Vessel Response Plans Oil, Section 155.1065 (c).

Title 33, Chapter I, Subchapter O, Part 155, Subpart D, Section 155.1050, Response Plan Development and Evaluation Criteria for Vessels Carrying Groups I through IV Petroleum Oil as a Primary Cargo.

Title 40, Chapter I, Subchapter J, Part 300, Subpart J, Section 300.910, Authorization of Use.

Transport Canada, *A Review of Canada's Ship-Source Oil Spill Preparedness and Response Regime: Setting the Course for the Future*, Tanker Safety Panel Secretariat, 2013.

U.S. Army Corps of Engineers, Waterborne Commerce of the United States, 2020, Part 3–Waterways and Harbors: Great Lakes, 2022, p. xv. As of December 6, 2023: https://publibrary.planusace.us/document/9d97f5a7-d394-4878-bda4-e0fc9908e458

USCG—*See* U.S. Coast Guard.

U.S. Coast Guard, National Pollution Funds Center, "Oil Pollution Act of 1990 (OPA)," webpage, undated. As of November 4, 2022: https://www.uscg.mil/Mariners/National-Pollution-Funds-Center/about_npfc/opa/

U.S. Coast Guard, Acquisition Directorate Research and Development Center, *Great Lakes Oil-In-Ice Demonstration 3: Final Report*, CG-D-8-13, revised October 2013.

U.S. Coast Guard, Argo After-Action Report, undated a, Not available to the general public.

U.S. Coast Guard, *M/V Harvest Spirit Marine Transportation Disruption After Action Report: 02 Dec 2020–04 Dec 2020*, provided to authors, undated b.

U.S. Coast Guard, NOSMAC PREP FSE 2022 After Action Report: 05 Apr 2022–17 May 2022, provided to authors, undated c.

U.S. Coast Guard, United States Coast Guard Response Resource Inventory PAV Report, spreadsheet provided to authors, undated c.

U.S. Coast Guard, *Great Lakes Area Contingency Plan*, USCG Ninth District, April 13, 2015. As of December 5, 2023: https://rrt5.org/Portals/0/docs/GreatLakesAreaContingencyPlan4-13-2015-Signed.pdf

U.S. Coast Guard, *Western Lake Superior Area Contingency Plan*, MSU Duluth, October 28, 2016.

U.S. Coast Guard, Eastern Great Lakes Area Contingency Plan Geographic Response Plan for Lake Ontario, 2017a.

U.S. Coast Guard, Eastern Great Lakes Area Contingency Plan Geographic Response Plan for Northeast Ohio, 2017b.

U.S. Coast Guard, Eastern Great Lakes Area Contingency Plan: Geographic Response Plan for Northwestern Pennsylvania, 2017c.

U.S. Coast Guard, Eastern Great Lakes Area Contingency Plan: Geographic Response Plan for St Lawrence Seaway, 2017d.

U.S. Coast Guard, Eastern Great Lakes Area Contingency Plan, Sector Buffalo, 2017e.

U.S. Coast Guard, Northern Michigan Area Contingency Plan, Sector Lake Michigan, 2017f.

U.S. Coast Guard, Northern Michigan Area Contingency Plan, Sector Sault Sainte Marie, 2017g.

U.S. Coast Guard, *Great Lakes Oil Spill Response and Cleanup Activities Assessment, Report to Congress*, November 21, 2018a.

U.S. Coast Guard, Manistique incident briefing ICS 201, provided to authors, September 20, 2018b.

U.S. Coast Guard, Sector SSM PREP FSE 2018 After Action Report: 19 Sep 2018–20 Sep 2018, provided to authors, February 15, 2019a.

U.S. Coast Guard, Northwest Ohio & Southeast Michigan Area Contingency Plan, Sector Detroit, May 13, 2019b.

U.S. Coast Guard, Coast Guard Sector Lake Michigan Area Contingency Plan, May 14, 2019c.

U.S. Coast Guard, "Great Lakes Icebreaker Acquisition Plan: Fiscal Year 2020 Report to Congress," June 17, 2020. As of May 5, 2023: https://www.dhs.gov/sites/default/files/publications/uscg_-_great_lakes_icebreaker_acquisition.pdf

U.S. Coast Guard, Guidelines for the U.S. Coast Guard Oil Spill Removal Organization Classification Program, 2021a.

U.S. Coast Guard, Northern Michigan Area Contingency Plan, 2021b.

U.S. Coast Guard, Acquisition Directorate Research and Development Center, *Freshwater In-situ Oil Burning*, CG-D-01-21, February 2021c. As of December 5, 2023: https://rrt5.org/Portals/0/PDFs/Freshwater%20In-Situ%20Oil%20Burning%20Report%20CG-D-01-21.pdf?ver=2021-04-19-085814-760

U.S. Coast Guard, *Guidelines for the U.S. Coast Guard Oil Spill Removal Organization Classification Program*, December 2021d. As of December 5, 2023: https://homeport.uscg.mil/Lists/Content/Attachments/55022/OSRO%20Guidelines%20-%20December%202021.pdf

U.S. Coast Guard Great Lakes Oil Spill Center of Excellence, Stakeholder Workshop, September 20, 2022.

U.S. Department of Homeland Security, *Department of Homeland Security Manual for the Operation of the Joint Requirements Integration and Management System*, DHS Instruction Manual 107-01-001-01, April 4, 2016.

U.S. Department of Transportation, Bureau of Transportation Statistics, *Fleet Composition of Rail Tank Cars Carrying Flammable Liquids: 2022 Report*, 2022. As of December 5, 2023: https://www.bts.gov/sites/bts.dot.gov/files/2022-09/BTS_Tank_Car_Report_To_Congress_2022_9-13-22_508.pdf

U.S. Energy Information Administration, "U.S. Energy Atlas," undated. As of December 5, 2023:
https://atlas.eia.gov

U.S. Energy Information Administration, "All Energy Infrastructure and Resources," web mapping application, updated April 20, 2023. As of December 5, 2023: https://atlas.eia.gov/apps/eia::all-energy-infrastructure-and-resources/explore

U.S. Environmental Protection Agency, National Response Center, database, undated.

U.S. Environmental Protection Agency, "Regional and Geographic Offices," webpage, updated January 17, 2023. As of December 5, 2023: https://www.epa.gov/aboutepa/regional-and-geographic-offices

U.S. Government Accountability Office, *Coast Guard: Improved Analysis of Vessel Response Plan Use Could Help Mitigate Marine Pollution Risk*, GAO 20-554, September 29, 2020.

U.S. Government Accountability Office, *Offshore Oil Spills: Additional Information is Needed to Better Understand the Environmental Tradeoffs of Using Chemical Dispersants*, GAO-22-104153, December 2021. As of December 5, 2023: https://www.gao.gov/assets/gao-22-104153.pdf

U.S. National Ice Center, "Great Lakes: Ice Charts," webpage, undated. As of June 2, 2023: https://usicecenter.gov/Products/GreatLakesHome

Wang, J., X. Bai, H. Hu, A. Clites, M. Colton, and B. Lofgren, "Temporal And Spatial Variability of Great Lakes Ice Cover, 1973–2010," *Journal Of Climate,* Vol. 25, No. 4, 2012.

Wang, J., J. Kessler, F. Hang, H. Hu, A.H. Clites, and P. Chu, "Great Lakes Ice Climatology Update of Winters 2012–2017: Seasonal Cycle, Interannual Variability, Decadal Variability, and Trend for the Period 1973–2017," NOAA Technical Memorandum, GLERL-170, December 12, 2017. As of December 5, 2023: http://www.glerl.noaa.gov/pubs/tech_reports/glerl-170/tm-170.pdf

Watkins, Robert K., and Alan A. Allen, "Remote Sensing Guide to Oil Spill Detection in Ice-Covered Waters," Arctic Response Technology, June 12, 2016. As of December 5, 2023: https://www.arcticresponsetechnology.org/wp-content/uploads/2017/09/Remote-Sensing-Guide-to-Oil-Spill-Detection-in-Ice-Covered-Waters.pdf

Wheeling, Kate, "What Causes Ecological Shifts?" Eos.org, October 3, 2017. As of December 4, 2023:
https://eos.org/research-spotlights/what-causes-ecological-shifts

Whitmer, Gretchen, "Governor Whitmer Takes Action to Shut Down the Line 5 Dual Pipelines Through the Straits of Mackinac After a Reasonable Transition Period to Protect the State's Energy Needs," press release, State of Michigan Office of the Governor, November 13, 2020. As of December 5, 2023:
https://www.michigan.gov/whitmer/news/press-releases/2020/11/13/governor-whitmer-takes-action-to-shut-down-the-line-5-dual-pipelines-through-the-straits-of-mackina

Wilkinson, J., C. J. Beegle-Krause, K-U. Evers, N. Hughes, A. Lewis, M. Reed, and P. Wadhams, "Oil Spill Response Capabilities and Technologies for Ice-Covered Arctic Marine Waters: A Review of Recent Developments and Established Practices," Ambio, Vol. 46, Supp. 3, December 2017. As of December 5, 2023:
https://www.ncbi.nlm.nih.gov/pmc/articles/PMC5673874/

Wirth, Anna Jean, Dulani Woods, Katherine Anania, Gary Cecchine, and Debra Knopman, Developing the Great Lakes National Center of Expertise for Oil Spill Preparedness and Response: An Opportunity to Reduce Risk and Impacts of Future Spills in Freshwater, Homeland Security Operational Analysis Center operated by the RAND Corporation, RR-A1222-1, 2021. As of September 29, 2023:
https://www.rand.org/pubs/research_reports/RRA1222-1.html